Micromachines for Biological Micromanipulation

Qingsong Xu

Micromachines for Biological Micromanipulation

Springer

Qingsong Xu
Department of Electromechanical
Engineering
University of Macau
Macau
China

ISBN 978-3-319-74620-3 ISBN 978-3-319-74621-0 (eBook)
https://doi.org/10.1007/978-3-319-74621-0

Library of Congress Control Number: 2017964244

Printed on acid-free paper

This Springer imprint is published by the registered company Springer International Publishing AG part of Springer Nature
The registered company address is: Gewerbestrasse 11, 6330 Cham, Switzerland

To my family

Preface

Micromachines (or microrobots) refer to miniature devices (typical size in micrometer and millimeter scales) which are powered to produce motions or forces for performing an intended action. According to performance requirement, a micromachine can incorporate structural elements, actuators, sensors, control components, and interfaces to achieve a specific objective. Such devices have been extensively developed for various applications including biology, medical sciences, chemistry, energy, environmental sciences.

Currently, biological micromanipulation emerges as an important approach in biomedical engineering. It concerns the manipulation of biological entities involving positioning, gripping, injecting, cutting, and fusion. In particular, the single cell (typical size around 10–500 μm) acts as the basic component of life, as it is the smallest unit of biological things. Therefore, biological cell micromanipulation has attracted extensive interests from both academia and industry in the past two decades.

Micromachines are ideal tools for micromanipulation of biological entities due to the merit of size matching and capability of producing/measuring motion and force in microscale. This book is focused on the enabling technologies in the development of micromachines dedicated to biological cell micromanipulation. It covers both mechanical designs and control designs toward precise and delicate micromanipulation tasks. The book proposes new mechanism designs of microforce sensor, microjector, microgripper, and microsyringe along with experimental verifications. The book also presents new position and force control of the microinjection systems dedicated to biological cell microinjection. A comprehensive treatment of the subject matter is provided in a manner amenable to readers ranging from researchers to engineers, by supplying detailed experimental verifications of the developed devices.

This book is composed of ten chapters. The book begins with an introduction to biological micromanipulation and micromachines and offers a brief survey of recent development of micromachines for biological micromanipulation applications. Chapter 2 presents a comprehensive review of cell microinjection systems, where the current development progress and remaining challenges are outlined.

Chapter 3 deals with the development of microforce sensor for the use in biological cell microinjection. A new force sensor is designed and fabricated based on polyvinylidene fluoride (PVDF) and macrofiber composite (MFC) films. Chapter 4 proposes the development of a microinjector driven by piezoelectric actuator for cell microinjection tasks. The superiority of force control over position control in survival rate improvement is demonstrated by experimental study. A smooth position/force switching control is introduced to alleviate the injury caused to the cells during the transition between the position and force control processes. Chapter 5 develops a constant-force microinjector which can produce a constant force output without a force sensor and controller. It provides a low cost and robust solution to eliminate the dependence on microforce sensor and controller.

The idea of constant-force mechanism is further extended to design two microgrippers in Chaps. 6 and 7, respectively. The effectiveness of the built constant-force microgrippers is validated by performing microgripping of biological cells and other micro-objects. Afterward, in Chap. 8, a force-sensing microelectromechanical systems (MEMS) microgripper is designed and fabricated for bio-sample grasp. The single sensor is able to provide the two-axis force sensing in two orthogonal directions, which contributes to the compact design and cost reduction of the microgripper in terms of fabrication and hardware components.

Microsyringe pump is a popular device in biological micromanipulation. Chapter 9 reports the design and development of a flexure-based compliant microsyringe pump which is driven by a piezoelectric stack actuator. It can deliver the liquid in tiny tube with a resolution in sub-nanoliter level. As a comprehensive application, Chap. 10 describes the development of a microinjection system with integrated dual cameras and force sensor. Experimental study on microinjection of zebrafish embryos with visual servo control and force control has been performed to improve the success rate of microinjection and survival rate of the injected cells. Experimental results demonstrate the effectiveness of the custom-built microinjection system.

This book provides the state-of-the-art emerging techniques to the development of compliant micromachines for biological cell micromanipulation. It covers the topics of mechanism design, actuation and sensing, motion and force control, and experimental study. Detailed examples of their implementations are provided. Readers can expect to learn how to design and develop new micromachine devices to achieve the microinjection and microgripping manipulation of biological cells.

The author would like to acknowledge the National Natural Science Foundation of China (under Grant 51575545) and the Science and Technology Development Fund (FDCT) of Macao (under Grant 090/2015/A3 and 143/2016/A) for co-funding the projects. The authors are also grateful for the help provided by Editor Marta Moldvai and Editor Brian P. Halm from Springer Verlag.

Macau, China
November 2017

Qingsong Xu

Contents

Chapter 1
Introduction

Abstract This chapter provides an introduction of biological micromanipulation using various types of micromachines. A brief survey of both tethered and untethered micromachines with different working principles is given. As a subset of microelectromechanical systems (MEMS), lab-on-a-chip micromachines based on microfluidics devices have been presented. Force sensing and control techniques in biological micromanipulation have also outlined for a safe and reliable operation.

1.1 Biological Micromanipulation

Nowadays, biological micromanipulation emerges as an important approach in biomedical engineering. It concerns the operation of biological entities involving positioning, gripping, injecting, cutting, and fusion, etc. In particular, the single cell (typically, around 10–500 μm size) acts as the basic component of life, as it is the smallest unit of biological things. Therefore, biological cell micromanipulation has gained extensive interests from both academia and industry in the past two decades.

Biological cell micromanipulation involves the operation of probing, injecting, gripping, cutting, etc. for biological cells. Traditionally, biological cells are manipulated manually by an operator using the visual information provided by an optical microscope. For example, cell microinjection is a crucial manipulation for DNA therapy, intracytoplasmic sperm injection (ICSI), drug delivery, and so on. It can be operated manually by two hands of human operator under a microscope. While one hand immobiles the cell using a glass pipette with suction force, the other hand penetrates the cell embryo with a sharp pipette and then injects exogenous materials into the cell. However, manual operation suffers from low efficiency, low success rate, and low repeatability. Moreover, the long-time operation will cause fatigue to the human operator.

Alternatively, robotic micromanipulation system allows the realization of precision positioning, gripping, and assembly operation of micro-objects by integrating actuation, sensing, and control of the microrobotic system. Such technology enables the dynamic exploration of objects in micro/nanoworld, and scale new height for the development on new material, information technology, manufacturing equipment,

Q. Xu, *Micromachines for Biological Micromanipulation*,
https://doi.org/10.1007/978-3-319-74621-0_1

biomedical science, and so on. The realization of robotic micromanipulation system requires the incorporation of various micromachines which function as micromanipulation tools with actuation and sensing capabilities. Such micromachines involve microprobe, microinjector, microgripper, etc.

Generally, microprobe is an efficient micromachine to move and separate biological cells. Equipped with force-sensing capability, it is useful for the characterization of mechanical property of cells. Microinjector is an indispensable micromachine in cell microinjection manipulation. An ideal microinjector is expected to bring the lowest injury to the cells, which facilitates the achievement of a high survival rate of the injected cells for subsequent biological study. For this purpose, microinjector with force sensing and control paves a promising way toward reliable cell microinjection. Microgripper functions as an important micromachine in pick–transport–place manipulation of biological cells. Force-sensing microgrippers are capable of regulating the grasp force for a safe micromanipulation. Mounted on a micropositioner, the force-sensing microgripper is a power tool dedicated to biological micromanipulation including cell separation, cell sorting, cell stiffness measurement.

In the early film Fantastic Voyage (1966), the author describes a team of scientists and submarine that are shrunken to microscopic size and injected into the bloodstream of a near-death scientist with a small crew. The mission of the voyagers is to relieve a blood clot caused by an assassination attempt. The film imagines a micromachine that can operate in human body. The development of such a tiny autonomous micromachine is still a challenging goal nowadays.

1.2 Tethered Micromachines for Bio-micromanipulation

Tethered type of micromachines involve mechanical microprobes, microinjector, and micrograppers, etc., for bio-micromanipulation as illustrated in Fig. 1.1. During the micromanipulation process, direct physical contact is established between the micromachines and biological samples.

Scanning probe microscopy (SPM) produces the image for surfaces by using a physical probe which scans over the specimen. SPM was founded in 1981 by the

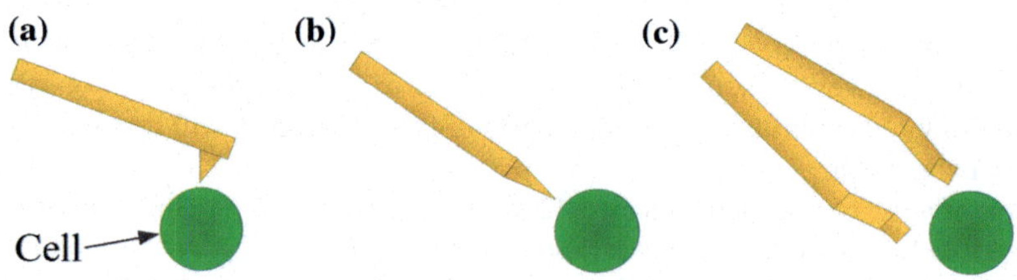

Fig. 1.1 Typical tethered micromachines for biological cell micromanipulation. **a** Microprobe; **b** microinjector; **c** microgripper

invention of the scanning tunneling microscope (STM). Later, since the invention of the atomic force microscopy (AFM) by Binnig, Quate, and Gerber in 1986 [9], AFM has been widely applied to the study in biological filed. In particular, as a typical microprobe, the cantilever of an AFM has been adopted for the manipulation, observation, and characterization of living cells [1, 6, 10, 19, 40]. Such applications have been enabled by the AFM owing to its dominant advantages in terms of high-resolution spatial imaging and the capability of force measurement. A recent survey of research progress in quantifying the mechanical properties of single living cells using AFM is presented in the literature [42]. Other types of SPM, e.g., piezoresponse force microscopy (PFM) [36], have also been explored for biological micromanipulation.

Cell microinjection is a kind of micromanipulation which introduces foreign materials (e.g., DNA, RNAi, sperm, protein, toxins, and drug compounds) into living cells. It has been extensively applied to genetics, transgenics, molecular biology, drug discovery, reproductive studies, and other biomedical areas [15, 49, 69, 73]. A custom-built automated robotic microinjection system is shown in Fig. 1.2. Usually, such manipulation is conducted using a microinjector (e.g., glass micropipette) with sharp tip. In order to reduce the deformation/damage of the cells and to increase the survival rate of the injected cells, the microinjector is usually driven by actuators will rapid response speed and high resolution of positioning, such as piezoelectric actuator (PZT). Piezo-driven microinjector was first introduced into cell injection by Kimura [55], where the microinjector was driven by a serial of piezopules to inject mouse oocytes. Since then, lots of remarkable research works have been conducted on piezo-driven cell microinjector [20, 21, 23, 29, 31, 35]. Majority of existing piezo-driven microinjectors operate based on the high-frequency vibration of PZTs. The tip is driven to vibrate in axial or lateral direction to penetrate the cell walls. Nevertheless, it has been argued that the high-frequency vibration may change the original status and cause adverse effects on the cell [84]. Some recent work on the development of nonvibration piezo-driven microinjector has been carried out in the literature [67, 68]. Piezo-driven devices exhibit nonlinear phenomena due to the hysteresis and

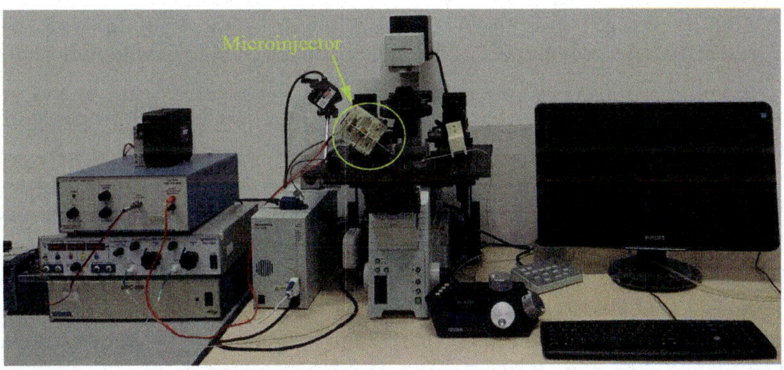

Fig. 1.2 Photo of an automated robotic cell microinjection system with inverted microscope and controllers for the injector

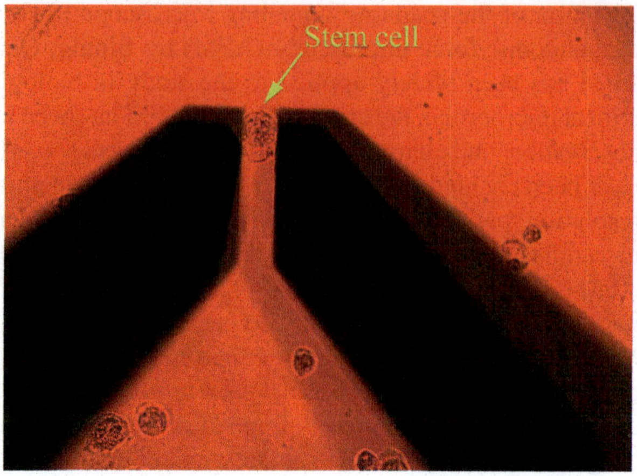

Fig. 1.3 Gripping of human mesenchymal stem cell (25 μm diameter) using a microgripper

drift effects. Precision motion control design for piezo-driven devices can be referred to the literature, e.g., [77]. In addition, the microinjector can also be adopted as a microprobe to deform the cell. The deformation of the cell is recorded to establish an analytical model to describe the cell deformation in cell mechanics study [62].

As a typical micromanipulation, microgripping is employed to realize the grasp or pick-and-place operation of a biological sample. As a contact-type micromachine, the gripper is usually designed based on flexure or compliant mechanism, which enables the elimination of friction, clearance, and backlash in motion transmission process. More details about the design of compliant mechanisms with large motion range can be referred to the book [76]. In the literature, various microgrippers have been presented with different actuators, such as electrothermal, electrostatic, shape memory alloy, piezoelectric, electromagnetic actuators. In addition to the integrated gripper structure design, chopstick-like microgripper can also be constructed by combining two individual micromanipulators for cell grasping and releasing operations [8]. The state-of-the-art reviews of the development on microgrippers have been reported in the literature [5, 34, 79]. Additionally, soft actuators (such as fluid power actuator [33, 83], electroactive polymer [22, 44]) have been adopted to develop microgripper for biological micromanipulation. Equipped with displacement and force sensors, the microgripper has been utilized to characterize the mechanical property of biological sample [37, 48, 75, 80]. For instance, force-sensing microgripper can be employed for stiffness characterization of biological cells to discriminate normal and cancer cells. Moreover, the microgripper also provides a promising approach to separate interested single cell from a group of cells (see Fig. 1.3), which offers a more convenient way alternative to conventional method based on microfluidic chips.

Cell cutting is an important manipulation for single cell analysis. For the cutting of biological cells, a microknife with sharp edge is required. Conventional diamond and glass knives cannot meet the requirement due to a large edge angle (over 20°). In the literature, the microknives fabricated from carbon nanotube (CNT) [59] and AFM cantilever [56] have been reported for single cell cutting.

Moreover, microelectromechanical systems (MEMS) have been recognized as ideal tools for biological cell micromanipulation and characterization, owing to their unique properties including size matching to single cells and capability of producing and measuring motion and force in microscale. A recent review of MEMS-based micromachines for mechanical manipulation and characterization of cells is presented in the literature [50].

1.3 Untethered Micromachines for Bio-micromanipulation

Untethered type of micromachines employ wireless types of drives based on optical, magnetic, and ultrasonic principles for bio-micromanipulation. During the micromanipulation process, contact or noncontact is made between the micromachines and biological samples. Unlike tethered micromachines, the micro-object is manipulated wirelessly with the help of produced field force.

Optical tweezers (also called "single-beam gradient force trap") are scientific instruments which use a highly focused laser beam to provide an attractive or repulsive force (typically in the order of piconewtons). Similar to a tweezer, it can be used to physically hold and move microscopic dielectric objects. Optical traps are capable of the manipulation and detection of sub-nanometer displacements for sub-micron dielectric particles. Due to this reason, optical tweezers have been widely adopted in biological applications, including trapping viruses and bacteria, manipulating cellular structures, and measuring forces of molecular motors and biological molecules such as DNA and proteins. For example, optical tweezers have been utilized for mechanical characterization of human red blood cells [63], indirect pushing-based automated micromanipulation of biological cells [64], and optical manipulation of multiple groups of micro-objects [26].

Magnetic tweezers are scientific instruments which are used for the manipulation and characterization of biomolecules or polymers. A magnetic tweezers apparatus consists of magnetic microparticles, which can be manipulated using an external magnetic field. The position of the magnetic particles is determined by a microscope with camera. These apparatus exert forces (typically in the order of pico- to nanonewtons) and torques to individual molecules or groups of molecules. Magnetic tweezers are commonly adopted to study the mechanical properties of biological macromolecules like DNA or proteins, the rheology of soft matter, and the force-regulated processes in living cells. Due to their simple architecture, magnetic tweezers are popular biophysical micromachine for micromanipulation and force measurement at the molecular level [24, 38, 46]. Recently, a 3D magnetic tweezer system has been reported for intraembryonic magnetic navigation and force application by introducing a single magnetic bead into a mouse embryo via robotic microinjection [70].

Magnetic mobile micromachines (also called microrobots) are devices which are driven by external magnetic field for diverse micromanipulation applications. These micromachines are usually fabricated into different shapes with magnetic materials, and their 2D or 3D mobility is enabled by varying the magnetic field. For example, helical-shaped magnetic microswimmers have been extensively developed toward biomedical applications, such as minimally invasive surgery, cell manipulation and analysis, and targeted therapy [52, 54]. Gripper-like magnetic micromachines have been proposed for the assembly of 3D structures, which have potential applications in the construction of medical devices inside the human body [16]. Magnetic intraocular microprobe has been created for measuring localized viscoelasticity of the vitreous body in vitrectomy surgery [53]. Magnetic microparticle has been fabricated for noncontact micromanipulation toward applications such as high-speed cell sorting and high-speed microassembly of artificial components [18]. A survey of recent progress on magnetically driven mircomachines is addressed in [11], which summarizes different designs including helical swimmers, flexible swimmers, surface walkers, and their applications in the fields of biomedicine or environmental remediation. For example, it has been reported that a sperm can be captured by a magnetic helical swimmer and delivered to the oocyte. Moreover, a comprehensive review of biomedical applications of magnetic actuation mobile mircomachines is presented in the recent literature [60].

Acoustic tweezers are instruments that are able to control the movement of objects by sound waves. In the standing acoustic field, objects can be moved to special regions by the experienced acoustic radiation force. Acoustic waves have been proven safe to biological objects, which make it an ideal tool for biomedical applications. More recently, acoustic tweezers have been popularly applied in such applications as cell separation, cell trapping, cell cutting, single cell manipulation, and so on [25, 32, 39, 45].

Other types of untethered biological micromanipulation tools include dielectrophoresis [66], microbubble [12].

1.4 Lab-on-a-Chip Micromachines for Bio-micromanipulation

A lab-on-a-chip (LOC) is a miniature device that integrates one or several laboratory functions on a single plate with size of millimeters to a few square centimeters. As a subset of MEMS, LOC devices commonly use microfluidics to precisely control and manipulate fluids that are geometrically constrained to a small (typically sub-millimeter) scale. Such low volumes of fluids are manipulated to achieve multiplexing, automation, and high-throughput screening. As microfluidic systems, LOC micromachines have been employed in biotechnology, clinical diagnosis, pharmacy, nature science, tissue engineering, nanomedicine, etc. [17].

Currently, the most popular technology for the fabrication of LOC micromachines for cell biological application is based on the soft-lithography of poly-di-methyl-siloxane (PDMS). PDMS is an elastomer which can be made into microfluidic devices through simple molding procedures [65]. In the literature, extensive works have been devoted to the development of LOC micromachines for the single cell analysis [82] and manipulation of biological cells based on magnetic, optical, mechanical, and electrical principles [81]. LOC micromachines have been fabricated for cell migration study [41] and microfluidic cytometer [13].

A LOC micromachine has been reported for the automation of cloning process in an all-in-one microfluidic chip by integrating the functions of cell cutting, sorting, coupling, and fusion [78]. Moreover, the LOC micromachine can be unified with a tethered or untethered micromachine for specific micromanipulation tasks. For instance, LOC micromachines have been designed to transport cells in automated biological cell microinjection applications [14, 61]. LOC micromachines have also been integrated with magnetic mobile micromachines for multiple-channel cell sorting tasks [27]. Dielectrophoresis-based LOC micromachines have been exploited for biological cell separation [7, 47].

In practice, different types of micromachines can be integrated together to enable a particular bio-micromanipulation application. Besides, soft microactuators [28] and soft micromachines [30] have been developed toward biological micromanipulation.

1.5 Microscopes for Biological Micromanipulation

Owing to the tiny size of the biological cells, optical microscopes are popular apparatus to observe and monitor the biological micromanipulation process. Both brightfield and fluorescence microscopes have been adopted. In addition, optical microscopes with equipped camera enable the implementation of visual servo motion control with the help of image processing technique. With a proper calibration, the visual feedback can also provide the force measurement by monitoring the deformation of the cell. For example, visual-based force measurement has been realized for the applications in single cell manipulation and characterization in the literature [43].

Due to diffraction barrier, the resolution of optical microscope is limited to approximately 0.2 μm. To obtain a higher resolution, a SPM can be adopted. In addition, the scanning electron microscope (SEM) provides nanometer resolution in real-time imaging with a large scanning area, which enables the integration of micromachines inside a vacuum chamber for simultaneous imaging and manipulation of samples [85]. However, conventional SEMs work in high-vacuum conditions, which prohibits the observation of water-containing samples, e.g., living biological cells. Recently, environmental scanning electron microscopes (ESEMs) have been developed to overcome this limitation. It allows the observation of liquid-phase materials, such as biological cells without metal coating and other electrically insulating materials [58]. In the literature, the simultaneous observation and real-time micromanipulation of biological samples for cell property characterization and surgery has been reported [2–4, 57].

1.6 Microforce Sensing and Feedback Control

Unlike macroworld, the micro-objects under manipulation are prone to be damaged if the contact force is not faithfully detected and precisely controlled. In biological micromanipulation, the safe manipulation requires micromachines with microforce-sensing capability. Although the force-sensing methods in macroworld have been well developed, these methods are not all suitable for detecting micronewton or nanonewton forces in microworld. Usually, the forces used for manipulating biological samples are too small to be felt by human operators. For example, tiny blood vessels are too delicate to be felt by the hand of the surgeon. In ophthalmological surgery, microforce sensing is required by the surgeon to exert image-guided intervention using optical coherence tomography [51]. Thus, there is a demanding tendency to improve the perception of forces in microworld.

Commonly, the size of force sensors should match with that of the sensing target. Macroforce sensors are generally not suitable for detecting microforce signals. The development of high-resolution, high-precision microforce sensors is of great significance for safe and reliable micromanipulation of biological entities. Available microforce sensors work based on strain gauge, piezoresistive, capacitive, piezomagnetic, optical, vision, electroactive principles, etc. Figure 1.4 depicts the photo a piezoresistive microforce sensor supporting a micropipette for microinjection [67]. The piezoresistive sensor is glued on a cantilever made of polyvinylidene fluoride (PVDF) film. The sensor works based on piezoresistive effects of silicon and measures the change in resistance caused by the exerted force. A state-of-the-art survey of recent research progress on microforce-sensing techniques has been presented in the literature [71].

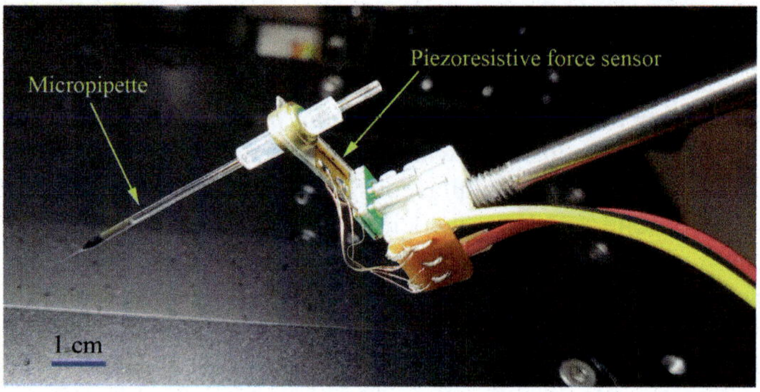

Fig. 1.4 A piezoresistive microforce sensor supporting a micropipette

Only a few of commercial microforce sensors are available on the market, including FT-S Microforce-Sensing Probe produced by FemtoTools AG[1] and AE801 series sensor elements provided by Kronex Technologies Corporation.[2]

It has been demonstrated that the microforce sensor provides a feature signal, which is helpful to determine whether the cell is injected or not in microinjection application [72]. Moreover, the capability of microforce sensing enables the implementation of force feedback control, which is desired in many micromanipulation applications to regulate the force precisely. In microinjection, a smooth transition during position and force switching control is important to reduce the damage to the cells [68]. In the literature, robotic microinjection of zebrafish embryos with force control has been presented to mimic human-injection operation [74]. It has been shown that the force control enables a higher survival rate for the injected cells than conventional position control [67].

1.7 Conclusion

This chapter presents a brief survey on the field of micromachines for biological cell micromanipulation. The micromachines are categorized into three general types including tethered micromachines, untethered micromachines, and LOC micromachines which involve the hybrid micromachines of different types.

Tethered micromachines consist of microprobe (e.g., SPM scanning cantilever), microinjector, microgripper, microknife, and so on. They are used to execute probing, injecting, griping, and cutting manipulation of biological cells. Such micromanipulation can also be implemented by untethered micromachines involving optical tweezers, magnetic tweezers, magnetic mobile microrobots, acoustic tweezers, dielectrophoresis, microbubble, etc. LOC micromachines commonly work based on microfluidics. They are usually integrated with tethered or untethered micromachines to carry out complex manipulation tasks of biological cells. Different kinds of microscopes for observing the micromanipulation process are introduced. Microscopes also offer the option of visual servo control for the micromanipulation system. Moreover, to guarantee a safe micromanipulation without damaging the cells, the promising solution of microforce sensing and control is also addressed. This book mainly deals with the development of tethered micromachines with microforce sensing and control to achieve safe and reliable biological cell micromanipulation.

---

[1] http://www.femtotools.com/.

[2] http://www.kronex.com/.

References

1. Afrin, R., Zohora, U.S., Uehara, H., Watanabe-Nakayama, T., Ikai, A.: Atomic force microscopy for cellular level manipulation: imaging intracellular structures and DNA delivery through a membrane hole. J. Mol. Recognit. **22**(5), 363–372 (2009)
2. Ahmad, M.R., Nakajima, M., Kojima, M., Kojima, S., Homma, M., Fukuda, T.: Nanofork for single cells adhesion measurement via ESEM-nanomanipulator system. IEEE Trans. NanoBiosci. **11**(1), 70–78 (2012)
3. Ahmad, M.R., Nakajima, M., Kojima, S., Homma, M., Fukuda, T.: Nanoindentation methods to measure viscoelastic properties of single cells using sharp, flat, and buckling tips inside ESEM. IEEE Trans. NanoBiosci. **9**(1), 12–23 (2010)
4. Ahmad, M.R., Nakajima, M., Kojima, S., Homma, M., Fukuda, T.: Buckling nanoneedle for characterizing single cells mechanics inside environmental SEM. IEEE Trans. Nanotechnol. **10**(2), 226–236 (2011)
5. Ai, W., Xu, Q.: Overview of flexure-based compliant microgrippers. Adv. Rob. Res. Int. J. **1**(1), 1–19 (2014)
6. Alonso, J.L., Goldmann, W.H.: Feeling the forces: atomic force microscopy in cell biology. Life Sci. **72**(23), 2553–2560 (2003)
7. Alshareef, M., Metrakos, N., Perez, E.J., Azer, F., Yang, F., Yang, X., Wang, G.: Separation of tumor cells with dielectrophoresis-based microfluidic chip. Biomicrofluidics **7**(1), 011803 (2013)
8. Avci, E., Hattori, T., Kamiyama, K., Kojima, M., Horade, M., Mae, Y., Arai, T.: Piezo-actuated parallel mechanism for biological cell release at high speed. Biomed. Microdev. **17**(5), 98 (2015)
9. Binnig, G., Quate, C.F., Gerber, C.: Atomic force microscope. Phys. Rev. Lett. **56**, 930–933 (1986)
10. Chang, K.C., Chiang, Y.W., Yang, C.H., Liou, J.W.: Atomic force microscopy in biology and biomedicine. Tzu Chi Med. J. **24**(4), 162–169 (2012)
11. Chen, X.Z., Hoop, M., Mushtaq, F., Siringil, E., Hu, C., Nelson, B.J., Pane, S.: Recent developments in magnetically driven micro- and nanorobots. Appl. Mater. Today **9**, 37–48 (2017)
12. Chen, Y., Lee, S.: Manipulation of biological objects using acoustic bubbles: a review. Integr. Comp. Biol. **54**(6), 959–968 (2014)
13. Chen, Y., Nawaz, A.A., Zhao, Y., Huang, P.H., McCoy, J.P., Levine, S., Wang, L., Huang, T.J.: Standing surface acoustic wave (SSAW)-based microfluidic cytometer. Lab Chip **14**(5), 916–923 (2014)
14. Chow, Y.T., Chen, S., Liu, C., Liu, C., Li, L., Kong, C.W.M., Cheng, S.H., Li, R.A., Sun, D.: A high-throughput automated microinjection system for human cells with small size. IEEE/ASME Trans. Mechatron. **21**(2), 838–850 (2016)
15. Desmaële, D., Boukallel, M., Régnier, S.: A planar structure sensitive to out-of-plane forces for the force-controlled injection of suspended and adherent cells. In: Proceedings of the 2011 Annual International Conference of the IEEE Engineering in Medicine and Biology Society (EMBC), pp. 8420–8423 (2011)
16. Diller, E., Sitti, M.: Three-dimensional programmable assembly by untethered magnetic robotic micro-grippers. Adv. Funct. Mater. **24**, 4397–4404 (2014)
17. Dittrich, P., Manz, A.: Lab-on-a-chip: microfluidics in drug discovery. Nat. Rev. Drug Discov. **5**, 210–218 (2006)
18. Dkhil, M., Bolopion, A., Regnier, S., Gauthier, M.: Analysis of the influence of inertia for non-contact micromanipulation. J. Micro Bio Rob. **13**, 15–26 (2017)
19. Dufrene, Y.F.: Atomic force microscopy in microbiology: new structural and functional insights into the microbial cell surface. MBio **5**(3), e01363–14 (2014)
20. Ediz, K., Olgac, N.: Effect of mercury column on the microdynamics of the piezo-driven pipettes. J. Biomech. Eng. **127**(3), 531–535 (2005)
21. Fan, M., Feng, Z., Agca, Y., Critser, J.K.: Vibration study of the piezodriven pipettes immersed in viscous liquids. J. Appl. Phys. **100**(7), 074701 (2006)

22. Feng, G.H., Yen, S.C.: Electroactive polymer actuated gripper enhanced with iron oxide nanoparticles and water supply mechanism for millimeter-sized fish roe manipulation. In: Proceedings of 2017 IEEE Electron Devices Technology and Manufacturing Conference (EDTM), pp. 216–218 (2017)
23. Gan, Y., Chen, Z.: A study of the zona piercing process in piezodriven intracytoplasmic sperm injection. J. Appl. Phys. **104**(4), 044702 (2008)
24. Gosse, C., Croquette, V.: Magnetic tweezers: micromanipulation and force measurement at the molecular level. Biophys. J. **82**(6), 3314–3329 (2002)
25. Guo, F., Mao, Z., Chen, Y., Xie, Z., Lata, J.P., Li, P., Ren, L., Liu, J., Yang, J., Dao, M., Suresh, S., Huang, T.J.: Three-dimensional manipulation of single cells using surface acoustic waves. Proc. Natl. Acad. Sci. **113**(6), 1522–1527 (2016)
26. Haghighi, R., Cheah, C.C.: Optical manipulation of multiple groups of microobjects using robotic tweezers. IEEE Trans. Rob. **32**(2), 275–285 (2016)
27. Hagiwara, M., Kawahara, T., Iijima, T., Arai, F.: High-speed magnetic microrobot actuation in a microfluidic chip by a fine V-groove surface. IEEE Trans. Rob. **29**(2), 363–372 (2013)
28. Hines, L., Petersen, K., Lum, G.Z., Sitti, M.: Soft actuators for small-scale robotics. Adv. Mater. **29**, 1603483 (2017)
29. Huang, H., Mills, J.K., Lu, C., Sun, D.: A universal piezo-driven ultrasonic cell microinjection system. Biomed. Microdev. **13**(4), 743–752 (2011)
30. Huang, H.W., Sakar, M.S., Petruska, A.J., Pane, S., Nelson, B.J.: Soft micromachines with programmable motility and morphology. Nat. Commun. **7**, 12263 (2016)
31. Huang, T., Kimura, Y., Yanagimachi, R.: The use of piezo micromanipulation for intracytoplasmic sperm injection of human oocytes. J. Assist. Reprod. Genet. **13**(4), 320–328 (1996)
32. Jeong, H., Li, T., Gianchandani, Y.B., Park, J.: Ultrasound-assited micro-knife for cellular scale surgery. In: Proceedings of IEEE 27th International Conference on Micro Electro Mechanical Systems (MEMS), pp. 885–888 (2014)
33. Jeong, O.C., Konishi, S.: All PDMS pneumatic microfinger with bidirectional motion and its application. J. Microlectromech. Syst. **15**(4), 896–903 (2006)
34. Jia, Y., Xu, Q.: MEMS microgripper actuators and sensors: the state-of-the-art survey. Recent Pat. Mech. Eng. **6**(2), 132–142 (2013)
35. Johnson, W., Dai, C., Liu, J., Wang, X., Luu, D.K., Zhang, Z., Ru, C., Zhou, C., Tan, M., Pu, H., Xie, S., Peng, Y., Luo, J., Sun, Y.: A flexure-guided piezo drill for penetrating the zona pellucida of mammalian oocytes. IEEE Trans. Biomed. Eng. (2017). https://doi.org/10.1109/TBME.2017.2713302
36. Kalinin, S.V., Rar, A., Jesse, S.: A decade of piezoresponse force microscopy: progress, challenges, and opportunities. IEEE Trans. Ultrason. Ferroelectr. Freq. Control **53**(12), 2226–2252 (2006)
37. Kim, K., Liu, X., Zhang, Y., Cheng, J., Wu, X.Y., Sun, Y.: Mechanical characterization of polymeric microcapsules using a force-feedback MEMS microgripper. In: Proceedings of the 30th Annual International Conference of the IEEE Engineering in Medicine and Biology Society, pp. 1845–1848. Vancouver, BC, Canada (2008)
38. Kim, K., Saleh, O.A.: A high-resolution magnetic tweezer for single-molecule measurements. Nucleic Acids Res. **37**(20), e136 (2009)
39. Lam, K.H., Li, Y., Li, Y., Lim, H.G., Zhou, Q., Shung, K.K.: Multifunctional single beam acoustic tweezer for non-invasive cell/organism manipulation and tissue imaging. Sci. Rep. **6**, 37554 (2016)
40. Li, G., Xi, N., Yu, M., Salem, F., Wang, D.H., Li, J.: Manipulation of living cells by atomic force microscopy. In: Proceedings of the 2003 IEEE/ASME International Conference on Advanced Intelligent Mechatronics (AIM 2003), pp. 862–867 (2003)
41. Li, J., Lin, F.: Microfluidic devices for studying chemotaxis and electrotaxis. Trends Cell Biol. **21**(8), 489–497 (2011)
42. Li, M., Liu, L., Xi, N., Wang, Y.: Research progress in quantifying the mechanical properties of single living cells using atomic force microscopy. Chin. Sci. Bull. **59**(31), 4020–4029 (2014)

43. Liu, X., Sun, Y., Wang, W., Lansdorp, B.M.: Vision-based cellular force measurement using an elastic microfabricated device. J. Micromech. Microeng. **17**(7), 1281–1288 (2007)
44. Lumia, R., Shahinpoor, M.: IPMC microgripper research and development. J. Phys. Conf. Ser. **127**(1), 012002 (2008)
45. Marx, V.: Biophysics: using sound to move cells. Nat. Methods **12**(1), 41–44 (2015)
46. McAndrew, C.P., Tyson, C., Zischkau, J., Mehl, P., Tuma, P.L., Pegg, I.L., Sarkar, A.: Simple horizontal magnetic tweezers for micromanipulation of single DNA molecules and DNA–protein complexes. BioTechniques **60**(1), 21–27 (2016)
47. Millet, L.J., Park, K., Watkins, N.N., Hsia, K.J., Bashir, R.: Separating beads and cells in multi-channel microfluidic devices using dielectrophoresis and laminar flow. J. Visualized Exp. **48**, 2545 (2011)
48. Muntwyler, S., Kratochvil, B.E., Beyeler, F., Nelson, B.J.: Two-axis micro-tensile tester chip for measuring plant cell mechanics. In: Proceedings of 2010 IEEE Sensors Conference, pp. 2451–2454 (2010)
49. Nakayama, T., Fujiwara, H., Tastumi, K., Fujita, K., Higuchi, T., Mori, T.: A new assisted hatching technique using a piezo-micromanipulator. Fertil. Steril. **69**(4), 784–788 (1998)
50. Pan, P., Wang, W., Ru, C., Sun, Y., Liu, X.: MEMS-based platforms for mechanical manipulation and characterization of cells. J. Micromech. Microeng. **27**, 123003 (2017)
51. Payne, C.J., Rafii-Tari, H., Marcus, H.J., Guang-Zhong, Y.: Hand-held microsurgical forceps with force-feedback for micromanipulation. In: Proceedings of 2014 IEEE International Conference on Robotics and Automation (ICRA), pp. 284–289 (2014)
52. Peyer, K.E., Tottori, S., Qiu, F., Zhang, L., Nelson, B.J.: Magnetic helical micromachines. Chem. Eur. J. **19**, 28–38 (2013)
53. Pokki, J., Ergeneman, O., Sevim, S., Enzmann, V., Torun, H., Nelson, B.J.: Measuring localized viscoelasticity of the vitreous body using intraocular microprobes. Biomed. Microdev. **17**(5), 85 (2015)
54. Qiu, F., Nelson, B.J.: Magnetic helical micro- and nanorobots: toward their biomedical applications. Engineering **1**(1), 021–026 (2015)
55. Ron-El, R., Liu, J., Nagy, Z., Joris, H., Van den Abbeel, E., Van Steirteghem, A.: Intracytoplasmic sperm injection in the mouse. Hum. Reprod. **10**(11), 2831–2834 (1995)
56. Shen, Y., Nakajima, M., Kojima, S., Homma, M., Fukuda, T.: Nano knife fabrication and calibration for single cell cutting inside environmental SEM. In: Proceedings of 2010 International Symposium on Micro-NanoMechatronics and Human Science, pp. 316–320 (2010)
57. Shen, Y., Nakajima, M., Zhang, Z., Fukuda, T.: Dynamic force characterization microscopy based on integrated nanorobotic afm and sem system for detachment process study. IEEE/ASME Trans. Mechatron. **20**(6), 3009–3017 (2015)
58. Shi, C., Luu, D.K., Yang, Q., Liu, J., Chen, J., Ru, C., Xie, S., Luo, J., Ge, J., Sun, Y.: Recent advances in nanorobotic manipulation inside scanning electron microscopes. Microsyst. Nanoeng. **2**, 16024 (2016)
59. Singh, G., Rice, P., Mahajan, R.L., McIntosh, J.R.: Fabrication and characterization of a carbon nanotube-based nanoknife. Nanotechnology **20**(9), 095701 (2009)
60. Sitti, M., Ceylan, H., Hu, W., Giltinan, J., Turan, M., Yim, S., Diller, E.: Biomedical applications of untethered mobile milli/microrobots. Proc. IEEE **103**(2), 205–224 (2015)
61. Song, P., Dong, X., Liu, X.: A microfluidic device for automated, high-speed microinjection of *Caenorhabditis elegans*. Biomicrofluidics **10**(1), 011912 (2016)
62. Tan, Y., Sun, D., Huang, W., Cheng, S.H.: Mechanical modeling of biological cells in microinjection. IEEE Trans. NanoBiosci. **7**(4), 257–266 (2008)
63. Tan, Y., Sun, D., Wang, J., Huang, W.: Mechanical characterization of human red blood cells under different osmotic conditions by robotic manipulation with optical tweezers. IEEE Trans. Biomed. Eng. **57**(7), 1816–1825 (2010)
64. Thakur, A., Chowdhury, S., Svec, P., Wang, C., Losert, W., Gupta, S.K.: Indirect pushing based automated micromanipulation of biological cells using optical tweezers. Int. J. Rob. Res. **33**(8), 1098–1111 (2014)

65. Velve-Casquillas, G., Berre, M.L., Piel, M., Tran, P.T.: Microfluidic tools for cell biological research. Nano Today **5**, 28–47 (2010)
66. Wang, C.H., Lee, Y.H., Kuo, H.T., Liang, W.F., Li, W.J., Lee, G.B.: Dielectrophoretically-assisted electroporation using light-activated virtual microelectrodes for multiple DNA transfection. Lab Chip **14**(3), 592–601 (2014)
67. Wang, G., Xu, Q.: Design and development of a piezo-driven microinjection system with force feedback. Adv. Rob. **31**(23–24) (2017). https://doi.org/10.1080/01691864.2017.1362996
68. Wang, G., Xu, Q.: Design and precision position/force control of a piezo-driven microinjection system. IEEE/ASME Trans. Mechatron. **22**(4), 1744–1754 (2017)
69. Wang, W., Liu, X., Gelinas, D., Ciruna, B., Sun, Y.: A fully automated robotic system for microinjection of zebrafish embryos. PloS one **2**(9), e862 (2007)
70. Wang, X., Luo, M., Wu, H., Zhang, Z., Liu, J., Xu, Z., Johnson, W., Sun, Y.: A three-dimensional magnetic tweezer system for intraembryonic navigation and measurement. IEEE Trans. Rob. (2017). https://doi.org/10.1109/TRO.2017.2765673
71. Wei, Y., Xu, Q.: An overview of micro-force sensing techniques. Sens. Actuators A Phys. **234**, 359–374 (2015)
72. Wei, Y., Xu, Q.: Design of a PVDF-MFC force sensor for robot-assisted single cell microinjection. IEEE Sensors J. **17**(13), 3975–3982 (2017)
73. Xie, Y., Sun, D., Liu, C., Tse, H.Y., Cheng, S.H.: A force control approach to a robot-assisted cell microinjection system. Int. J. Rob. Res. **29**(9), 1222–1232 (2010)
74. Xie, Y., Sun, D., Tse, H.Y.G., Liu, C., Cheng, S.H.: Force sensing and manipulation strategy in robot-assisted microinjection on zebrafish embryos. IEEE/ASME Trans. Mechatron. **16**(6), 1002–1010 (2011)
75. Xu, Q.: Design, fabrication, and testing of an MEMS microgripper with dual-axis force sensor. IEEE Sensors J. **15**(10), 6017–6026 (2015)
76. Xu, Q.: Design and Implementation of Large-Range Compliant Micropositioning Systems. Wiley, New York (2016)
77. Xu, Q., Tan, K.K.: Advanced Control of Piezoelectric Micro-/Nano-Positioning Systems. Springer, Berlin (2015)
78. Yamanishi, Y., Sakuma, S., Iyanagi, T., Arai, F., Arai, T., Hasegawa, A., Tanikawa, T., Ichikawa, A., Satoh, O., Nakayama, A., Aso, H., Goto, M., Takahashi, S., Matsukawa, K.: Design and fabrication of all-in-one unified microfluidic chip for automation of embryonic cell manipulation. J. Rob. Mechatron. **22**, 371–379 (2010)
79. Yang, S., Xu, Q.: A review on actuation and sensing techniques for MEMS-based microgrippers. J. Micro-Bio Rob. **13**(1), 1–14 (2017)
80. Yang, S., Xu, Q., Nan, Z.: Design and development of a dual-axis force sensing MEMS microgripper. J. Mech. Rob. **9**(6), 061011 (2017)
81. Yi, C., Li, C.W., Ji, S., Yang, M.: Microfluidics technology for manipulation and analysis of biological cells. Anal. Chim. Acta **560**, 1–23 (2006)
82. Yin, H., Marshall, D.: Microfluidics for single cell analysis. Curr. Opin. Biotechnol. **23**, 110–119 (2012)
83. Yoshida, K., Tsukamoto, N., wan Kim, J., Yokota, S.: A study on a soft microgripper using mems-based divided electrode type flexible electro-rheological valves. Mechatronics **29**, 103–109 (2015)
84. Yu, Y., Ding, C., Wang, E., Chen, X., Li, X., Zhao, C., Fan, Y., Wang, L., Beaujean, N., Zhou, Q.: Piezo-assisted nuclear transfer affects cloning efficiency and may cause apoptosis. Reproduction **133**, 947–954 (2007)
85. Zimmermann, S., Tiemerding, T., Fatikow, S.: Automated robotic manipulation of individual colloidal particles using vision-based control. IEEE/ASME Trans. Mechatron. **20**(5), 2031–2038 (2015)

Chapter 2
Review of Microinjection Systems

Abstract Cell microinjection plays an important role in genetics, transgenics, molecular biology, drug discovery, reproductive studies, and other biomedical fields. Robotic cell microinjection has been popularly applied due to its high precision, high repeatability, and high throughput. In this chapter, the state-of-the-art research on microinjection of both adherent cells and suspended cells with microforce-sensing techniques is reviewed. The challenges and promising methods in automating the cell microinjection process are discussed.

2.1 Introduction

2.1.1 The Role of Cell Microinjection

The purpose of cell microinjection is to introduce small volume of foreign materials into living cells [36, 37]. Since its introduction in the first half of the last century, injection of foreign materials (e.g., DNA, RNAi, protein, sperm, toxins, and drug compounds) into single living cells has been broadly applied in genetics, transgenics, molecular biology or drug discovery, reproductive studies, and other biomedical areas [14, 65, 99, 108].

In genetics, the genetic sequence of human DNA has been confirmed, and a total of about 30,000 genes were identified in the first phase of the Human Genome Project [69]. Confirming the gene sequence and identification of the genes is only the first step. The functions of each gene and produced proteins should be identified in the next step. Cell injection plays a crucial role in the acceleration of the second step in the Human Genome Project by transferring microfabricated DNA microarrays which prominently increase the quantity of experimental data [118]. An important technique for identifying gene functions is RNA interference (RNAi) through microinjection [10, 115]. RNAi was firstly stated in 1998 by manually injecting double-stranded RNA (dsRNA) [115]. dsRNA expresses endogenous enzymes to recognize, break the corresponding messenger RNA (mRNA), and hence silence the gene function. RNAi is regarded as a potential therapeutic strategy too [93]. A change in phenotype demonstrates the function of the silenced gene.

Q. Xu, *Micromachines for Biological Micromanipulation*,
https://doi.org/10.1007/978-3-319-74621-0_2

In transgenics, the injection of DNA is utilized to produce transgenic zebrafish lines [99]. Injection of mRNA is applied to overexpress gene products in zebrafish embryos. In addition, the loss-of-gene-function studies need the injection of antisense morpholino-modified oligonucleotides (morpholinos or MOs) to specifically prevent RNA splicing and/or translation [99]. The transgenic organisms have been generated for the last 30 years. The transgenic animal, e.g., the creation of "knockout mice," acts as a research tool to better understand the functional consequences of gene expression or deletion [69].

In molecular biology or drug discovery, molecule screening at single cell level, which is crucial in drug discovery and molecular biology, demands the target molecules to be transferred into single cells. Then, the cellular function-targeted molecules can directly control the cell development as well as their functions.

In the reproductive studies, intracytoplasmic sperm injection (ICSI) has gradually been the main method to overcome intractable male-factor infertility. It has become a routine process for lots of in vitro procedures [39].

It has been proved that cell injection has significant meanings to human beings [7]. Biological cells are irregular in shape and easily deformable, and they may be lightly destroyed during the manipulation and injection. At present, this greatly precise process is mostly done manually [109]. The development of suitable human-system interaction is still a challenging research area [19].

2.1.2 Conventional Manual Cell Microinjection

In biological research, manual microinjection is a traditional and extensive practice [99]. For instance, during the injection of zebrafish embryos, a human operator firstly identifies the embryos in a petri dish by using a microscope and then moves the tip of the micropipette toward the embryo carefully. The embryo is immobilized with another micropipette. At the moment when the micropipette tip initially touches the outer layer of the embryo, the operator will manually actuate the micromanipulator to generate a rapid thrust movement, leading to a preliminary penetration of the embryo by the micropipette. After this preliminary penetration, the operator will continue actuating the micromanipulator, until the micropipette tip penetrates into the embryo's yolk. Then, genetic materials can be injected into the embryo by a microsyringe, such as a pneumatic microsyringe [19]. During this manual microinjection process, the experience and skill of the operator play a vital role in realizing a successful injection.

Usually, it takes several months or one year to train an operator to become proficiency in completing such a task, that is a time-consuming process [19, 52, 75, 93, 115]. Despite the long training time, the success rate of a skilled worker can only reach around 15% due to the tough conditions (i.e., long-time concentration, patience, and causing fatigue) to achieve a successful injection [21, 39, 68, 95]. Even so, a successful injection also needs five crucial conditions, i.e., an accurate injection point, proper speed variation, suitable penetration trajectory, appropriate penetration force, and no contamination [52, 82].

Specifically, an accurate injection point, e.g., the center of a cell, is necessary to avoid micropipette slipping on the cell surface without penetrating (or missing) the nucleus after penetrating. Proper speed variation means that the speed of micropipette touching the outer layer of the cell accelerates to high enough speed to make the micropipette rapidly pierce the cell membrane, without inducing excessive deformation and destroying the membrane to obtain the integrity of the membrane. This is crucial in transgenesis task, because the successful integration of the genetic material into the genome is also needed in the nucleus [69]. After that, the speed should be decelerated to zero as soon as possible to avoid destroying the inner structure of cells. A suitable penetration trajectory will reduce the time that is required to finish an injection operation. It is particularly crucial for the scenarios where a number of cells need to be injected. For instance, in order to test the cellular responses to molecular targets, and to obtain the statistically significant data, thousands of cells require to be injected in a short time period (e.g., within 1.5 h after the fertilization) because of quick succession [99]. For instance, the injection of embryos with dsRNA is generally conducted within the first 60 min of embryonic development [115]. Proper penetration force prevents the cell from being destroyed by excessive force or hand tremor. The condition of no contamination requires the reduction or elimination of the participation of human. Obviously, it is hard to always satisfy all of the rigorous conditions at same time for a human operator. Hence, low success rate, poor repeatability, and extended training time are resulted. Even more important, it is impossible to implement the regional (or organ specific) delivery within an embryo or fetus via manual operations [68]. It is practical only when treating small numbers of embryos and cells [118].

In order to overcome the problems intrinsic in manual injection (e.g., human fatigue, long training time, labor intensive, and poor repeatability), various cell injection methods have been developed to achieve a higher success rate.

2.1.3 Current Methods of Cell Microinjection

Several methods are available for delivering foreign materials into embryos or cells, such as chemical methods, vehicular methods (e.g., erythrocyte fusion and vesicle fusion) [55, 96], electrical method (e.g., electroporation [71]), and mechanical methods (e.g., microinjection, hyposmotic shock [6], sonication [88], and microprojectiles [46]).

Microinjection normally uses a fine micropipette for penetrating through a cell membrane and delivering liquid into the cell with a pressure pulse [28]. Microinjection is the standard method for injecting embryos and cells, because it can reproducibly and reliably introduce large numbers and specified volume of macromolecules to majority types of embryos and cells with high viability and function [118]. It is important that microinjection can achieve quantitative delivery of multiple components into the same cell at any specified time without introducing other potentially confounding compounds [56, 57]. It also should be noted that microinjection is a better choice than vectors, because vectors randomly integrate to the host DNA and can induce uncontrolled influence on the gene expression and cause phenotypical changes. In microinjection, the cells to be injected can be carefully selected and hence the injected cell can be marked, which can improve the success rate of injection and the reliability of study. Moreover, several cells (e.g., primary cells and stem cells) are hard to transfect via traditional methods [28]. Moreover, the same end-effector (i.e., a fine micropipette) can be utilized in different operations (e.g., electrophysiological measurements and cell isolation [28]). In contrast to other techniques, microinjection with a fine micropipette is the most effective approach in consideration of the issues including cell viability, cell damage, cell waste, effectiveness of introducing macromolecules, and keeping from concerning about phenotype alteration.

Because of two distinct biomanipulation tasks [22], the cells to be injected in microinjection can be classified into two types, i.e., adherent and suspended cells, which are presented in detail below.

2.2 Injection of Adherent Cells

Adherent cells grow at the bottom of a petri dish and form a fixed cell population. By contrast, suspended cells grow loosely, as shown in Fig. 2.1. Expect for blood cells and germ cells, all of other cells in human body are of adherent cells. Therefore, the research on adherent cells is very important in drug development and disease mechanism studies. Adherent cells are usually smaller in size (10–20 μm in diameter) than suspended cells (e.g., oocytes—the most frequently injected suspended cells). Because the size of the adherent cells is nearly 5–10 times smaller than that of oocytes, microinjection of adherent cells needs more accurate micromanipulators in terms of positioning and injection accuracy. In addition, as the adherent cells are small in size and they usually grow in population close to each other, they are hard to be detected. This imposes a high demand on the vision and other measurement systems. The small size of the cells also imposes requirement for very fine injection capillaries. Since the fine micropipettes (<1 μm) are demanded, it is very hard to

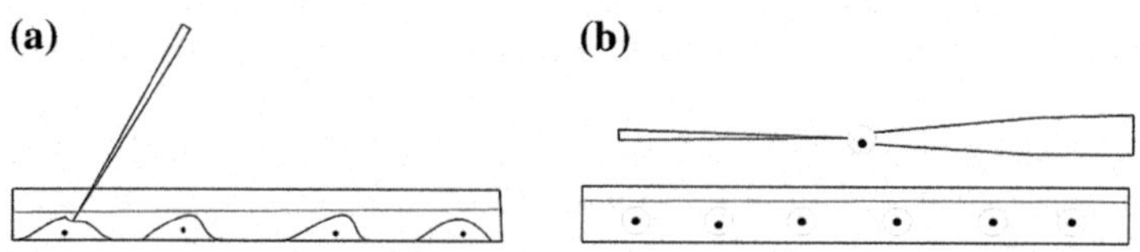

Fig. 2.1 Sketch diagram of microinjection of **a** adherent cells and **b** suspended cells

visually monitor the condition of the micropipette, contact with cell, correct injection depth, and so on [28]. In addition, owing to the fact that adherent cells are fixed at the bottom of petri dish, the holding pipette is not required.

Some commercial devices are also available, such as the products offered by Eppendorf, Kleindiek, Newport, and Narishige and Cellbiology Trading. The most popular commercial adherent cell injection device is AIS 2, that is produced by Narishige and Cellbiology Trading.[1] This partly automated cell injection device is primarily composed of an inverted microscope, two three-axis stages driven by stepping motors for positioning the cells and controlling the micropipette movement, a piezo-driven axial injector, a pneumatic microinjector, a video system with a CCD camera and a monitor, and the associated software for controlling the cell injection process. By incorporating the functions of auto-focusing, identification of cells and injector pipette, positioning control with image-based visual serving and calibrated injection height control, this injection device can produce a high success rate of injection. The software commands the injection process precisely and enables a fast axial injection with a speed of 1500 cells per hour. In addition, retrieving the injected cells and multiple injections into the same cell are also achieved by the system [22].

Although some systems have already partly automated, all the devices currently need intensive manual work. Because of the participation of the operator, the quantity of cells which can be injected in a prescribed time interval is limited. It is an issue when plenty of cells have to be injected or when microinjection is adopted to create stable transfected cell lines [74]. Moreover, the operator's skill imposes a huge affect on the success rate. Another disadvantage is the lack for method to diagnose the condition of micropipette, such as clogging and small breakages. The drawbacks of the individual devices result in a low output, a low success rate, and decreased reliability of the results [7, 28].

Some researchers have devoted to improve the automation degree and success rate by adding force sensor. For instance, Lukkari et al. and Kallio et al. have developed an impedance sensor to detect the contact between the cell and inject micropipette, a broken micropipette, a clogged micropipette, an aged measurement electrode, and faulty injection solution [28, 53]. Meister et al. used atomic force microscopy (AFM) to achieve force-controlled injection of adherent cells [59]. Desmaele et al. developed a force sensor with planar structure to sense out-of-plane forces, and living cells can be placed on the planar sensor [14].

To conclude, the development of an automated robotic cell injection system for single adherent cell is a very challenging work. Due to the scaling effect, the relationships between physical quantities alter in the microworld. For instance, the gravity is less regnant than van der Waals forces and electrostatic forces. The uncertainties caused by the scaling effect make the operation more difficult. In addition, the uncertainties are induced by biology call for sufficient robustness for the cell injection system. Moreover, each type of cell exhibits its own specific properties, and the state of cell population changes over the time, which increases the difficulty in successful cultivation, detection, and injection of adherent cells [28].

[1] http://www.ais2.com/.

2.3 Injection of Suspended Cells

In the field of cell microinjection study, there are three types of most popular suspended cells, i.e., Drosophila embryo, zebrafish oocytes, and mouse oocytes.

2.3.1 Drosophila Melanogaster Embryo

The Drosophila melanogaster (fruit fly) embryo has relatively short reproductive cycle, small genome, and possibility in the cure of human diseases. It is one of the most popular organisms in biological research, medical study, developmental biology, and genetics. It also has been applied in studying the wiring of human brain and nervous system. Genetic modification of Drosophila embryos has aroused the interest from both scientific research and medical industry for finding biological mechanisms to treat diseases. The possibility to improve human health based on the research related to drosophila embryo has been verified by the award of 1995 Nobel Prize in Physiology or Medicine for the discoveries related to the genetic control of early embryonic development [76]. In that research, the fruit fly was taken as the test model. This organism is popular in genetics, because human and Drosophilas are similar in genetics. The principles obtained from fruit fly are suitable for higher organisms involving human.

The Drosophila genome can also offer important information about human genes which are homologous in the Drosophila melanogaster [72]. In particular, several human diseases are aroused by mutations in genes analogous to genes discovered in Drosophila. Specifically, approximately 61% of known human disease genes have an identifiable match in the genetic code of Drosophila, and 50% of Drosophila protein sequences exhibit mammalian analogs [73]. Drosophila has been utilized as a genetic model for a number of human diseases containing neurodegenerative disorders, e.g., Parkinson's disease, Alzheimer's disease, and Huntington's disease.

Moreover, Drosophila has been used to explore the mechanisms related to immunity, diabetes, cancer, and drug abuse. Additionally, the gene function can be determined from the loss-of-function phenotype or the overexpression phenotype. The function of human genes could be studied through introducing them into Drosophila by transposable elements. For instance, Goodman et al. [97] used Drosophila to explore the wiring of the brain and nervous system, leading to an improved understanding of how the human brain develops. Hence, the integration of the Drosophila genome with the well-developed genetic methods in the Drosophila system will give rise to crucial discoveries for human medicine and development in detecting, treating, and eradicating diseases in humans. In order to carry out the research on Drosophila genome, one of the most significant processes is the injection of materials which affect the composition of a cell (or an organism). Microinjection can generate Drosophila with new characteristics through integrating transgene or dsRNA into the DNA of Drosophila. For example, in RNAi microinjection tests, 100–200 Drosophila embryos need to be injected with 60 pl of dsRNA during the first 60 min of their development to assess one gene [118]. This enables us to determine which genes are vital for the development of the organism and which organs are influenced [76].

2.3.2 *Zebrafish Embryo*

The zebrafish (*Danio rerio*) embryo has been broadly adopted as a standard animal model for studying the development of vertebrate. Moreover, zebrafish has also been utilized as a model for understanding the human disease's pathogenic mechanisms and discovering drug [63, 66]. Microinjection of genetic materials into zebrafish embryo has been a routine process to test functions of the injected materials on the survival and development for embryos [24].

There are four distinct advantages making zebrafish hot research model. Firstly, zebrafish embryos are easily obtained and fertility. Secondly, they grow rapidly and the life cycle is short, i.e., approximately 12 weeks, which makes generic analysis easily. Thirdly, the characteristics of transparent and external fertilization make the development and change of cells be inspected easily [70]. Fourthly, solid organ malignancies developed in zebrafish are similar to human tumors, as the embryonic development of zebrafish is markedly analogous to that of humans [69].

It is worth noting that molecular and gene should be injected into early zebrafish embryos to analyze zebrafish embryo genesis [40]. As shown in Fig. 2.2, the diameter of zebrafish embryos is around 600–700 μm (without chorion) or 1.15–1.25 mm (with chorion), in which the cytoplasm and nucleus locate on the animal pole linked with a large mass of yolk. The diameter of the injecting micropipette tip for zebrafish embryos is about 6–10 μm [109]. The zebrafish embryo has four developmental phases, i.e., blastula phase, gastrula phase, pharyngula phase, and prehatching phase. Blastula embryos form three germ layers at 7.5 h post-fertilization and then enter the gastrula phase. The embryos are named pharyngula phase after 26-h post-fertilization, which shows distinct movement within the chorion. During the period of 26–48 h post-fertilization, the phase is named prehatching phase. The embryos hatch after 48-h post-fertilization when a protease enzyme is secreted to dissolve the chorion envelope [32]. The zebrafish can be collected according to the standard procedures of embryo preparation [104].

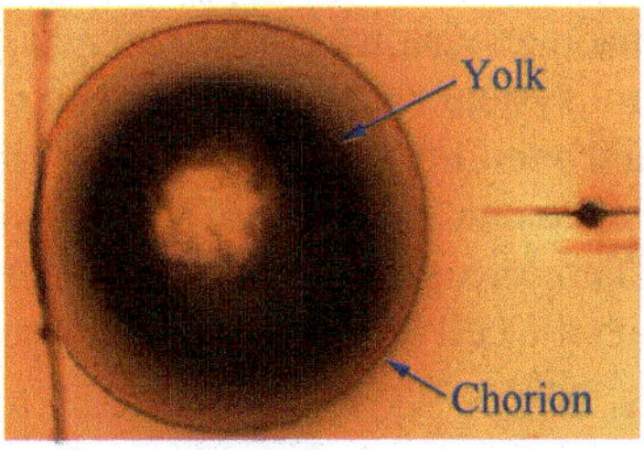

Fig. 2.2 Microscope image of a zebrafish embryo

2.3.3 *Mouse Embryo*

Microinjection of mouse embryos is significant for vitro fertilization, screening molecular targets related to the research of basic biology of embryo development (e.g., neutralizing antibodies, mitochondrial-associated recombinant proteins, morpholinos), and expression vectors for siRNA [47]. In addition, the mouse serves as a model organism, which is a primary animal for genetics and reproductive research.

Because of the freedom of movement of suspended cells in the nutrient solution, the suspended cell needs to be fixed by additional equipment, such as holding pipette, making injection procedure more complicated and time-consuming. Moreover, the collision between cells and manipulation instruments, or the motion of cells themselves, can easily result in the failure of this procedure. Even though there is a tough requirement on suspended cell injection, majority of the operations have been performed manually [22]. However, the manual method features low efficient, low repeatability, and low throughput, which cannot meet the need for testing genetic materials when multiple injection operations should be done in a certain time. On the contrary, the robotic cell injection system can realize that successful cell injection with high repeatability and accuracy in a certain time [107]. Therefore, the development of the robotic cell injection system has been the concentration of lots of researchers.

2.4 Robotic Cell Microinjection System

Robotic microinjection is a method which uses automation technology to introduce materials into individual living cells by a fine needle [108]. In comparison with manual microinjection, the dominant advantages of robotic cell microinjection involve more qualitative, more productive, more reliable, free from fatigue, and unparalleled repeatable. Robotic cell microinjection systems can operate in a greatly efficient and consistent way, and hence could notably improve the reproducibility and throughput of cell injection and even make it possible for new types of studies that cannot be achieved by traditional techniques [117]. Robotic cell injection is extremely preferable when a large number of cells need to be injected with abundant materials in a certain time.

Generally, a robot cell microinjection system includes piercing mechanisms with injection control loop, cell holder and micromanipulator (for precise position), machine vision and other nonvision sensors, user interface, and an environment control system for maintaining cell cultivation conditions (e.g., temperature, pH value, and humidity) [67]. By improving the level of automation of the robotic cell injection, the human involvement can be reduced and the cell injection speed can be increased. Therefore, scientists can concentrate on analyzing the results. Moreover, automated robotic cell injection can increase the reliability and accuracy and hence offer more reliable results [28].

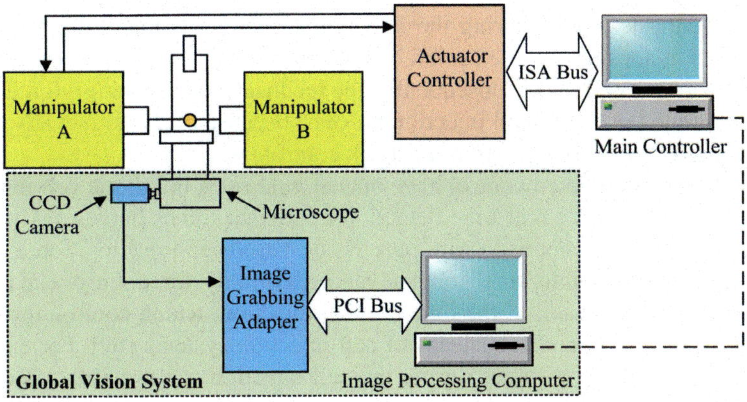

Fig. 2.3 Schematic diagram of an automated suspended cell microinjection system

Currently, some semi-automated or tele-operated robotic cell injection systems are commercially available, such as the products supplied by Eppendorf,[2] Narishige and Cellbiology Trading.[3] However, these products need greatly skilled operators to carry out suspended cell injection. In the literature, Li et al. [43, 44] have presented a representative automated suspended cell injection system, as illustrated in Fig. 2.3. The process of the automated suspended cell injection primarily contains four steps in the following.

- Firstly, it implements a visual-based search and recognition for the suspended cells, injecting pipette, and holding pipette by image processing.
- Secondly, the holding pipette is driven to hold the cells and then moved to the desired manipulation place.
- Thirdly, the injector pipette is guided to inject into cells and release materials at desired position.
- Fourthly, the injected cells are released in the culturing area.

The above process is repeated until all of the cells have been injected.

However, the injection process purely relies on position control, which is apparently not a reliable control process because of the following reasons. Firstly, if the injection force is too big, cells and the injection pipette may suffer from the possibility of being destroyed, because they are highly fragile. Secondly, there is no fast and accurate feedback during the procedure to make sure that the penetration is successful. For example, sometimes the injecting pipette may only slip over the surface of cells rather than penetrating into them. Though the vision feedback can be used to monitor the penetration procedure, the vision feedback is unpractical in recognizing the injecting pipette's tip correctly and quickly, due to the difficulty in detecting whether the tip is inside or outside the cell. Thirdly, the computational load

[2]http://www.eppendorf.com.

[3]http://www.narishige-group.com/.

of vision feedback control during the injection procedure is huge, due to the fact that multiple images need to be captured and processed in real time in the injection procedure for each cell [52]. Consequently, the feedback control using a force sensor is widely adopted in recent cell injection process, because of its distinct advantages over the pure vision-based position feedback control.

Force sensor and related control have played vital roles in robotic cell injection. Force sensor can provide real-time detection of contact forces between a injecting pipette and a cell. In comparison with pure vision-based cell injection, force sensor-based cell injection exhibits six distinct advantages. Firstly, force sensor can provide an accurate force feedback of the ongoing cell injection, which could improve the dexterity, success rate, and robustness of cell injection systems [69]. For example, the detection of injecting forces during the cell injection can precisely predict the penetration of cell membrane and hence initiate the subsequent material delivery [117]. Secondly, because the biological cells and injecting pipette are delicate, the quantification of contact forces between the injecting pipette and cells is helpful to prevent the cell and injecting pipette from excessive force [47]. Thirdly, precise measurement of contact forces is a necessary condition toward minimally invasive cell injections, which can improve the survive rate of injected cell. Experimental and theoretical data indicate that the extensional flow at the entrance of the injecting pipette is the main reason of biological cell death [1]. Fourthly, the sensing of cellular force is essential for understanding the biophysical properties of cell injury and membrane modeling issue [32]. Fifthly, force signals can also be helpful to detect the physical condition of the cell by assuming that a stiff membrane implies a weakened cell, which can reduce the waste of precious injecting materials [7]. Last but not least, force sensor-based cell injection can improve the speed and need not high-quality image processing equipment, causing cost reduction [85]. It should also be noted that the injection force is in μN–mN range [19].

2.5 Microforce Sensors for Cell Microinjection

Although the pure visual-based position control (visual servo) has been widely adopted to perform cell microinjection tasks [82, 99, 113], the measurement of contact forces during pipette injection plays a crucial role as it can be applied to provide force feedback for precise control of the needle penetration speed and strength [39]. Moreover, the visual-based position information is generally less effective than the force information. On the contrary, the direct force information can reflect the changes in the physical behavior of the cell (e.g., deformation or extent of penetration) more accurately and quickly [52]. Five popular microforce-sensing methods in cell injection are introduced in the following.

2.5.1 Vision-Based Force Sensors

Robotic cell injection is generally performed with the help of an optical microscope. Hence, visual feedback is the dominant sensing method in existing robotic cell injection systems [47]. Vision-based force sensors are used to determine cell injection forces by using image processing and an accurate cell model [19]. The forces are usually computed on the basis of the deformations of visually tracked flexible objects (e.g., cells, manipulation tools, or cell holders). And the measured geometrical information is used by a force estimation algorithm to provide the force sensing [34].

The vision-based force sensors have some distinct advantages. Firstly, vision-based force sensor can provide global forces feedback rather than local forces offered by contact force sensors, where the latter strongly limits the haptic rendering for the operator [39]. Secondly, the vision-based force sensors are most helpful when the force information is required and it is highly challenging or even impractical to use a force sensor [29]. For instance, very small-scaled and accurate force sensing for cell injection is more difficult, because the design of small force sensors requires to solve the challenging problems for cell injection, such as multiple degrees-of-freedom (DOF) microforce sensing with high resolution and accuracy, and high signal-to-noise ratio (SNR). In addition, the designed force sensors must maintain an adequate reliability and repeatability, because force sensors experience severe disturbances in cell injection due to the liquid surface tension and adhesion forces [18]. Thirdly, vision-based force sensors are able to obtain both vision and force information by a single vision equipment (e.g., charge-coupled device (CCD) or complementary metal-oxide semiconductor (CMOS) camera) under microscopic conditions, which does not increase the complexity of the cell injection system [21]. Therefore, new methods which can avoid the use of force sensors become very prevalent recently [34].

2.5.1.1 Image Processing for Vision-Based Force Sensor

Image processing is adopted to extract the required information as the input to the established model. The speed of the vision system is highly dependant on the speed of the localization and recognition procedure. The recognition of the cell and the micropipette can be realized by different approaches, such as matching method. Pattern matching and feature matching are two most frequently used matching methods. The pattern (template) matching implements direct localization on the original image by detecting the degree-of-similarity between an image and the template [34, 51]. The feature matching requires to extract features (e.g., active contour, brightness [115], or amplitude spectra) first and then implement matching in the feature space. Contour abstraction has been broadly investigated in image processing, and many approaches have been presented. The most frequently employed edge detection technique is the gradient-based Prewitt, Sobel, and Laplace detectors [13]. Other contour finding techniques like the second-derivative zero-crossing detector [54] or computational approach based on the Canny criteria [9] have been proposed. Nevertheless, owing to general image features such as noise, texture, image blur, or other anomalies like nonuniform scene illumination, edge finding techniques usually fail in providing satisfied results. For instance, the boundaries of adjacent cells may be wrongly recognized as broken edge fragments or might not be detected. In addition, the edge information cannot be detected when the edge only covers a few pixels. Eventually, these techniques generally need post-processing to get connected and closed contours.

As firstly presented by Kass et al. [30], the active contours (i.e., snakes) have been applied in many applications. It consists of the procedures of edge detection, shape modeling, segmentation, pattern recognition and object tracking [41]. Such technique always generates closed contours and is quite adapted to segment biological images. However, the existence of other objects (e.g., the holding pipette, the injecting pipette or the impurities in the medium) imposes disturbances and hinders the direct use of the snakes for contour tracking of the cell membrane deformations. In order to solve the problem, a series of preprocessing steps should be performed ahead. The first step includes erasing the holder pipette and the injecting micropipette from the image. The correlation-based template matching is used to locate the pipettes in real time [22]. The pipettes/membrane contact points define the boundaries of the holder pipette and the injecting micropipette segments. These contact points are acquired by detecting the gray level of the image along the pipettes' edges. The second step is composed of locating the impurities and removing them from the image [2–4].

2.5.1.2 Cell Model for Vision-Based Force Sensor

Physics-based model of the living cell has been built by combining together the cell geometric information obtained from image processing, a priori knowledge of cell mechanical properties, and a predefined coordinate system of the slave environment [34]. Then, the force can be computed from the model by providing the updated cell boundary condition. In the literature, mechanical cell models can be primarily divided into three categories including microscale continuum, energetic, and nanoscale structural types [39].

The first category of models assumes that the biological cell is equivalent to one or two phases continuum model rather than considering the molecular nanoscale mechanical properties [45]. A typical model of this category is shown in Fig. 2.4. The main advantages involve their ability to calculate the mechanical properties of cells and offer the details about the distribution for stresses and strains reacted on cells (e.g., zebrafish and medaka embryos) at different developmental stages [89, 90]. Sun et al. [86] have proposed a point-load model to estimate the cell injection force with cell deformation. In this model, some assumptions are presented to simplify the model, such as linear elastic biomembrane, uniform stress within the dimple, and zero residual stress in an initial planar circular area. However, such assumptions are not always true in reality.

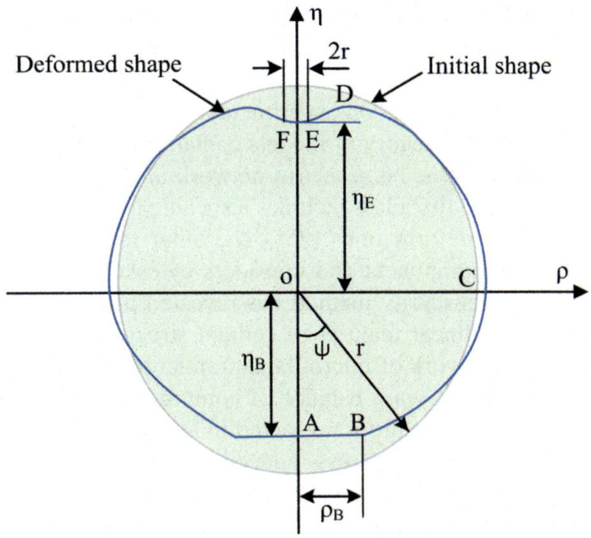

Fig. 2.4 Microscale continuum cell model for vision-based force sensor

Contour model [20] and active contour model [34] have also been presented to estimate the forces exerted on a linearly elastic object (i.e., cells) using the contour data, which are generated on the basis of visual data. In [4], a 3D nonlinear mass-spring-damper cell model is proposed to estimate the injection force. Nevertheless, such model can only provide limited accuracy and weak connection to biomechanics [34]. For instance, there is no mechanically related relationship between the model parameters and the cell properties. Furthermore, these parameters are calculated through offline finite element method (FEM) simulations by using finite element modeling, which is influenced by the network topology. In addition, Ref. [33] has reported a boundary element model (BEM) and a prior knowledge of the mechanical properties for the cell to predict the cell injection forces. Tan et al. [89, 90] have presented a mechanical model based on membrane theory with the quasi-static equilibrium equations, where the relationship between the cell injection forces and the deformation of cells is built. Huang and Sun et al. [21] have used the point-load model to estimate the cell injection force by using the displacement of inject pipette. Kim et al. [34] proposed another boundary element model to estimate the cell injection force, where the position update of an injecting pipette is used as the input for the model to estimate the injection forces so as to reduce the calculation load. Asgar et al. [5] have presented a 3D particle-based mechanical model, which is based on spring-damper model with multi-particle joints to estimate the injection forces. Nevertheless, considering the cell as a tensed balloon filled with molasses or jello, the major disadvantage of the continuum method lies in that it is not able to explain the molecular deformations and interactions within the cell.

The models of the second category consider the effects of various cytoskeleton structures into the entire energy budget of cell during contraction [26]. It is established based on the percolation theory and polymer physics models with large deformations.

The main advantage of the model is that it is independent in choosing coordinate system and the particular details of the cytoskeleton architecture, due to the energy is a scalar quantity. Nevertheless, it is hard to find an optimal physical correspondence between the model and experimental data.

The third category of models containing the tensegrity structures can be divided into two classes, i.e., spectrin-network model and cytoskeletal models. The model belonging to first class includes a specific microstructural network for spectrin cells with large deformations [25]. The latter regards the cytoskeleton as the primary structural component and considers cytoskeleton contractile forces as the central role. The tensegrity method has revealed many aspects of cell deformability containing nonlinear features of cellular structural behavior. The models consider the cell as a network of microfilament, microtubule, and actin, which distribute forces within the cell via a balance of compression and tension [58]. Those models can simulate many features presented in living cells during mechanical tests containing strain hardening, prestress-induced stiffening, and the effect of cell spreading on cell deformability [116]. A full *mechano-cell* model, which includes the cell membrane, the nuclear envelope, and actin filaments, has been proposed on the basis of the minimum of the elastic energy during deformation [94]. It is considered as a combination of various spring elements. Ladjal et al. [39] have proposed a simplified cytoskeleton tensegrity structure physically based on FEM model, which enables us to simulate the cell deformation through real-time simulation constraints. In the third model, the cell geometry and biomechanical subcomponents properties (e.g., biomembrane, cytoplasm, cytoskeleton, and nucleus) play the dominant role in simulation and modeling, as these factors influence the amount of cell membrane deformation, needle deviation, and interaction forces [4].

In addition to measure the deformation of cells, Liu et al. [47, 49] have measured the deflection of a polydimethylsiloxane (PDMS) post in a cell holder to estimate cell injection forces with a resolution of 3.7 nN. This method can be used to detect the injection forces of different cells, because the model is only related to the deflection and material parameters of the cell holder. Moreover, Karimirad et al. [29] have proposed an artificial neural network, which is trained by the existing relationship between cell parameters in images and force, to estimate the load for a spherical biological cell. This method is free from building a real mechanical cell model and relatively easy to perform. However, the inner structure of the cell in this model is ignored and the model is sensitive to parameters variation in biological cells. Furthermore, Ammi et al. [3] have also presented a 3D cell model constructed with virtual reality (VR) environment to improve the realism, where the force is estimated by the contour measurement based on vision technique.

It is also notable that the finite element model (FEM) needs a controlled slave environment to model the membrane [34]. The mass-spring model is commonly nonrealistic and greatly sensitive to the tuning of the model, e.g., in the spring constant of the mesh. Comparatively, boundary element method (BEM) is a numerical approach to solving the differential equations indicating an object model that computes the unknowns on the model boundary rather than its entire body. BEM uses less computation time than FEM, whereas FEM is difficult to construct and change the element mesh. By contrast, BEM is more suitable as it uses the boundary mesh rather than solid elements of FEM for the cell. Kim et al. [34] have adopted the cell edge information and known material properties to estimate the force.

However, vision-based force sensors will be affected by the inevitable parameters uncertainty in the dynamic cell model. Moreover, vision-based force sensors are greatly dependent on the availability of suitable cell model. In addition, the sampling frequency of the vision-based force sensors is highly lower than that of the contact force sensor, and thus the injection process cannot be controlled precisely. The performance of these sensors is limited by the small depth of field of the used microscope [22]. Furthermore, a priori knowledge of the cell properties should be obtained before estimating the forces [14]. The resolution of such force sensors is intrinsically limited by the optical components of the microscope, inhibiting application where the direct force feedback is needed.

2.5.2 Capacitive Force Sensors

Currently, majority of available capacitive force sensors applied in force detection of cell injection are based on microelectromechanical systems (MEMS). MEMS technologies are highly suitable for the fabrication of cell handling systems, because of the cost-effective mass-fabrication of custom-built injectors with integrated sensors [115]. MEMS sensors can be designed at the end-effector, causing a marked increase in accuracy, reproducibility, and reliability of the cell injection. MEMS sensors are able to provide qualitative and quantitative information on the cellular, sub-cellular and organism levels, that is useful to understand the fundamental elements of biological systems [84]. MEMS force sensors exhibit advantages of small size, wide bandwidth, and high sensitivity. Due to the significance of quantifying forces and their influence on the function and morphology of many biological structures, MEMS force sensors have played a crucial role in biological studies [84].

In these force sensor devices, very tiny deflections induced by exerted forces are transformed as detectable capacitance variations. An electronic circuit converts the capacitance changes into DC-voltage changes. MEMS capacitive force sensors can measure the forces ranging from pN and mN, which makes them be desirable for the force detection in cell injection. With their high performance and ability to perform measurements with multiple degrees-of-freedom, MEMS capacitive force sensors are powerful substitutes to other MEMS-based transducers, e.g., cantilever-based sensors [15].

Capacitive type of MEMS force sensors has five distinct advantages [84]. (a) They are able to measure a wide range of forces (from mN down to pN level), whereas the measurement range of other force-sensing methods is usually limited; (b) they are able to provide the force information of multiple axes; (c) they use the most direct method of force measurement rather than indirectly measuring approach; (d) they exhibit the merit of low noise, low power, and high sensitivity. Moreover, they are insensitivity to light, temperature, and humidity variations [62]; (e) they can be manufactured by batch microfabrication processes by using deep reactive ion etching (DRIE) on silicon-on-insulator (SOI) wafers [85], leading to cost effective.

The MEMS capacitive force sensors are designed as comb drives, i.e., an array of parallel-plate capacitors. There are two working modes in capacitive force sensor, i.e., transverse mode and lateral mode of comb drive movement. The transverse mode changes the capacitance by altering the gap size between the capacitor plates, while the lateral mode changes the capacitance by altering the overlapping area between the capacitor plates. The capacitive force sensor in transverse mode enables the design with higher resolution than that of lateral mode, at the expense of lack of linearity. It also should be noted that high stiffness is required for high-bandwidth measurements. During the design of capacitive force sensors in transverse mode, the differential configuration of comb drives is useful in creating linear relationship between the deflection and the sensor output signals for small deflections [62], as well as compensating for disturbances [81]. Compared with strain gauges and piezoelectric force sensors, capacitive force sensors are more stable and sensitive and exhibit no hysteresis [29]. As compared with an optical beam-based atomic force microscope, they also have stable output signals (with low drift) and high compact size [62].

In the literature, four representative capacitive force sensors are presented. Sun et al. [84, 85] have proposed a two-axis capacitive force sensor with high sensitivity (i.e., 0.01 and 0.24 μN) by using large overlapping area and transverse model. The sensor is applied in the characterization of the mechanical properties for mouse oocyte and embryo zona pellucida [86]. However, a constraint in motion displacement of the pipette is revealed in cell micromanipulation applications [29]. Muntwyler et al. [62] have proposed a three-axis microforce sensor with regulable force ranges (from ±20 to ±200 μN) through the readout electronics with different settings for wider applications. Moreover, a new microfabrication process with a double silicon-on-insulator (SOI) substrate has also been presented, leading to a major decrease in the fabrication complexity for multi-axis sensors [62]. Beyeler et al. [8] have proposed a six-axis MEMS capacitive force–torque sensor. Moreover, Xu [110] has proposed the sensor design using one transverse comb drive for sensing two perpendicular forces, namely gripping and interaction force sensing with minimum resolution of 0.61 μN, as depicted in Fig. 2.5.

2.5.3 *Optical-Based Force Sensors*

Optical force sensors commonly consist of a load transduction medium (e.g., microcantilever or grating, for experiencing the exerted force), a light source (e.g., light emitting diode (LED), laser, or halogen lamp), and a photo-detector (e.g., photodiode or CCD camera, for sensing ranks of illumination, refractive index, or spectrum of the light source, that vary with the deformation of load transduction medium [100]). Additionally, a circuitry is also required to transfer the detected variation as an inductive output signals.

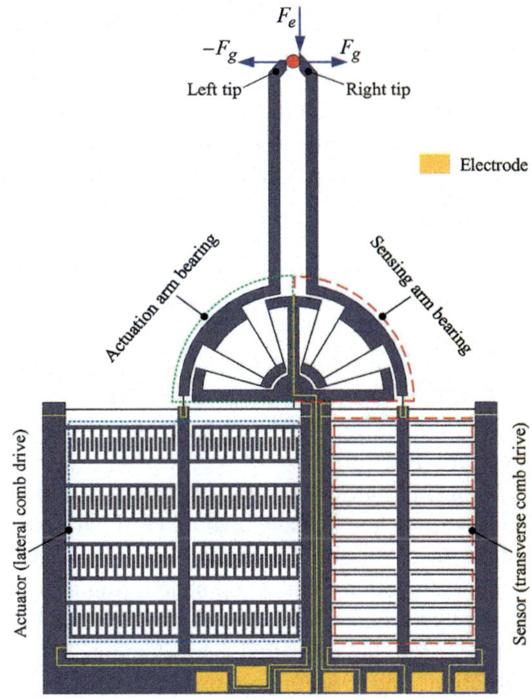

Fig. 2.5 A microgripper with a single capacitive-type force sensor for two-axis force sensing

Optical-based force sensors can offer the desired resolution and keep relatively large sensing ranges for microinjections. Moreover, such kind of force sensors provide a method to solve the conflict in designing the parameters about sensitivity and linearity [105]. The optical-based force sensors have excellent potential thanks to its extreme high resolution (down to nN scale) and the electromagnetic immunity [29]. Moreover, optical-based force sensor is an effective mean in noncontact force measurement, and atomic force microscope is one of the typical applications. Atomic force microscope is able to offer pN–nN force feedback for the cellular force measurement [47].

Cantilever-based optical force measurement has been widely used in atomic force microscopy (AFM). When a force is applied on the cantilever, the photodiode amplifies and senses the displacement of the cantilever. The force is computed by multiplying the optically sensed displacement by the known spring constant of the cantilever. The major error sources are the mechanical vibration, laser pointing stability, and shot noise for modulation frequencies over 10 kHz [60]. The system can be achieved with ultrasensitive and higher resolution by applying cantilevers with lower stiffness.

However, there are mainly four limitations which restrict the applications of cantilever-based optical force sensors in cellular force measurement. Firstly, AFM measurement requires a complex transmit–receive setup, which demands a high accuracy on optical alignment and adjustment [29]. In order to achieve a high accuracy, the surface of the cantilever must be adequately reflective. Secondly, it has a small force measurement range, which is limited by a small range of deflection of the photodiode. Thirdly, the force measurement may be inaccurate because of the refraction and reflection of the transmitted light through aqueous medium, in which the biological cells are cultured. Fourthly, AFM does not offer really simultaneous imaging and cell injection capabilities, which requires the end-effectors with microcantilever and a sharp tip [47].

Another traditional optical force sensor is based on laser traps (or optical tweezers) [106]. The resolution of laser trap can be generated in nN or sub-nN level. In laser trapping, the high-energy light (close to the UV spectrum) is required to achieve the high dissipation of visible light in aqueous solutions. It results in the possibility of damage to the cell and abnormalities in the cell's genetic materials, because the cells could absorb the high energy and heat. However, some researchers have claimed that such concerns can be solved by using the wavelengths in the near-infrared (IR) spectrum [83].

In the literature concerning cell injection applications, some representative designs are described here. Zhang et al. [119, 120] have proposed a 1D micrograting-based optical force sensor integrated with a silicon nitride injector to characterize the dynamic injection force when penetrating Drosophila embryos. The injector is supported by springs with known spring constant, and then the injection force can be deduced from the measured displacement. The displacement is measured by a high-resolution, linear, and miniaturized optical encoder, and the resolution is less than 1 μN with a range of 10 μN. Wiens et al. [105] have used an optimized Robert's mechanism to improve the linearity, dynamic range, and sensitivity of the design. In addition, Loh et al. [50] have adopted the automatic approach function of AFM to measure the cell injection force of 36 nN in a prescribed time.

2.5.4 Piezoresistive Force Sensors

As a type of strain sensor or strain gauge, piezoresistive force sensors work based on resistance change when a force is exerted. As compared with piezoelectric and capacitive microforce sensors, the piezoresistive force sensor generally offers more accurate and stable force signals in a large measurement range at the cost of a relatively low resolution in tens of mN level [35, 51]. The variation in resistance is usually measured through an electric circuit, such as a Wheatstone bridge. Due to the output of the piezoresistive sensor is analog, namely a voltage signal, its resolution highly depends on the noise level, thermal drift, power consumption and bandwidth of the amplifier, as well as the bridge configuration of the sensor [52]. Piezoresistive force sensors have also been used in AFM to measure the atomic force change at the nN level [27, 92].

In the literature, Lu et al. [52] have applied a commercial piezoresistive force sensor (model: AE801, from Kronex Technologies Corp.) to monitor the injection process of zebra fish embryo as shown in Fig. 2.6a, where the real-time force signal

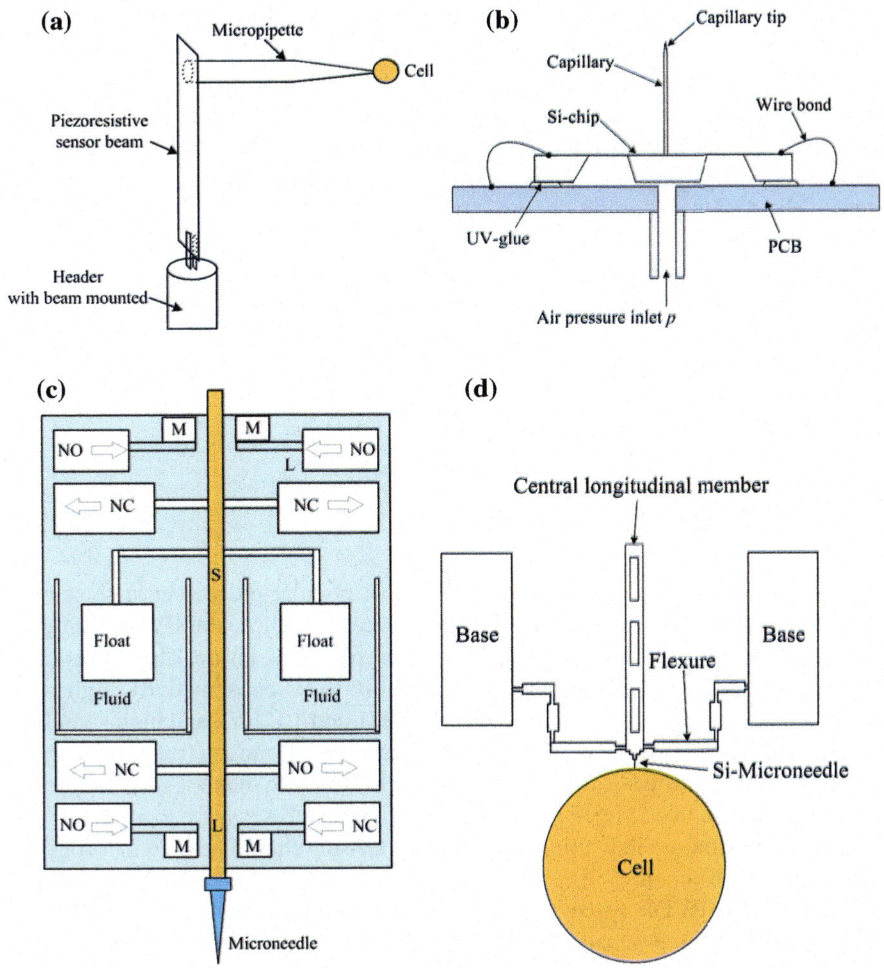

Fig. 2.6 Typical structures of piezoresistive force sensors. **a** A commercial piezoresistive force sensor (model AE801, from Kronex Technologies Corp.) is modified for use in penetration sensing, **b** MEMS piezoresistive microforce sensor, **c** silicon membrane-based piezoresistive force sensor, **d** silicon-structured piezoresistive force sensor

is applied to stop the injecting. The appearance of dramatic decrease in force signals indicates the penetration of the cell, because there is almost no resistance after a cell is penetrated. Shulev et al. [79] have proposed a MEMS piezoresistive microforce sensor with sub-pN sensitivity, as shown in Fig. 2.6b. It provides the ability to inject cell vertically and simultaneously with μN level injection force. Beutel [7] has reported a piezoresistive microforce sensor made of silicon membrane which offers a resolution of 120 μN, as shown in Fig. 2.6c. The sensor is able to self-calibrate and monitor the injecting pipette status, such as pipette break and cell sticking. The ability of monitoring pipette status greatly improves the reliability and throughput of the cell injection by providing online fault diagnostics of the injecting pipette. Stavrov [80] has proposed an axial piezoresistive microforce sensor fully made of silicon to monitor the cell injection forces, as shown in Fig. 2.6d. By tuning the amplification gain factor for the sensor's onboard electronics, the force measurement range varies from several tens of μN to several hundred mN with a resolution of nN to μN level, respectively. The main limitation of the previous works [52, 80] is that the material cannot be injected into cell with the structure that micropipette is directly attached to the sensor, which is not applicable in practice [29]. In addition, some piezoresistive force sensors need manual assembly, which may result in misalignments and significant errors in force measurement [35]. The resistance and the gauge factor of such sensors vary as temperature changes. Furthermore, they are sensitive to the bonding quality and bonding location.

2.5.5 *Piezoelectric Force Sensors*

Polyvinylidene fluoride (PVDF) film is the most popular piezoelectric sensor for force measurement in cell injection. When a force is applied, the PVDF film generates an electric charge based on the forward piezoelectric effect. The film is ideal for sensing the force of cell injection, because of its excellent sensitivity, high compliance, high-frequency bandwidth (between 0.001 and 10^9 Hz), and high signal-to-noise ratio [17]. Moreover, its properties of measurement range (from μN to mN), resolution (sub-μN to μN), and relatively simple structure make it suitable for the force measurement in cell injection [22]. Nevertheless, it is sensitive to acoustics and changes in temperature, indicating that the force sensors should be used in a constant temperature environment [29].

The available PVDF force sensors can be categorized into three types, i.e., cantilever-PVDF force sensor, simply supported PVDF force sensor, and fixed-guided PVDF force sensor, which are addressed as follows.

Cantilever-PVDF force sensor uses the PVDF film as a cantilever beam, while a needle is usually glued on the free end of the PVDF film and the other end is fixed on a manipulator. This kind of structure is mostly adopted to sense the injection force because of its extremely high sensitivity with a free end. Kim et al. [32] have used an unmodeled cantilever-PVDF force sensor to quantity the mechanical behavior of zebrafish embryo with 14.5-mN resolution. Shim et al. [78] have adopted the cantilever-PVDF structure to achieve μN level of resolution. Wejinya et al. have applied two pieces of PVDF films to build a 2D cantilever-PVDF microforce sensor to investigate the mechanical behavior of a fruit fly embryo [76, 102]. Pillarisetti et al. [69] have employed a cantilever-PVDF force sensor with resolution of μN to demonstrate the positive effect of force feedback fused with vision feedback in improving the success rate for cell injection, as shown in Fig. 2.7a. Interestingly, Shen et al. [77] have presented the concept of active force sensor to modify the cantilever-PVDF force sensor to obtain higher sensitivity with sub-μN resolution. Moreover, Shen

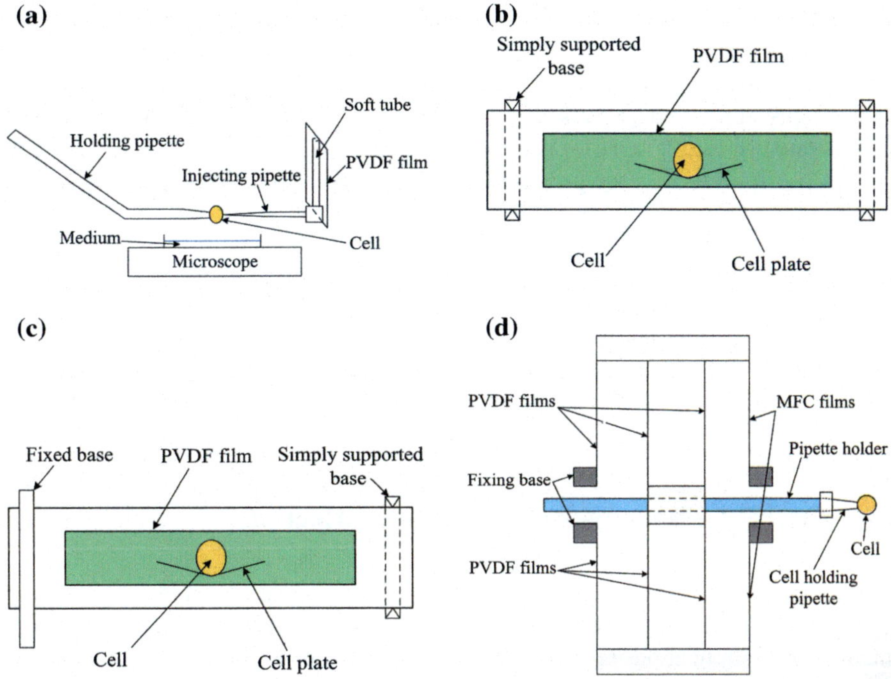

Fig. 2.7 Typical structures of PVDF force sensors. **a** Cantilever-PVDF force sensor, **b** simply-supported PVDF force sensor, **c** fixed-simply-supported PVDF force sensor, **d** fixed-guided beam-type PVDF force sensor

et al. [11, 76] have improved the performance of cantilever-PVDF force sensor by modeling the sensor and specially designed the related electric circuit with resolution of sub-μN for detecting the mechanical properties of living Drosophila embryos. Xi et al. have introduced a resonance frequency approach to measure the spring constant of cantilever-PVDF membrane and vision method to measure the deformation of cells to obtain the force information [103]. Huang et al. [22, 23] have proposed a cantilever-PVDF force sensor to identify the scenarios when the pipette contacts the cell and the cell is penetrated. However, in the aforementioned literature, the measured signals cannot be held still because of its inherent dynamic characteristic. Sun et al. [87] have presented an inverse-model signal processing method to develop a static PVDF microforce sensor.

Generally, the cantilever structure makes it hard to change the injecting pipette once broken. In addition, it is difficult to link the injecting pressure tube with the pipette needle, because the needles in majority of the aforementioned designs are directly bonded to the PVDF film [31, 32, 42]. Although injecting pressure tube is attached along the PVDF in the literature [69], the sensitivity is greatly affected by added weight and constraints.

In the simply-supported PVDF force sensor as shown in Fig. 2.7b, a PVDF film is adhesively pasted on the back of the supporting beam, while the cell plate is well placed on the center points of the beam [107–109]. A similar structure called fixed-simply-supported beam is also modified to construct a PVDF sensor in [114], as illustrated in Fig. 2.7c. However, the cells cannot be well immobilized and they can easily slip in these structures, which will decrease the efficiency, stability, and accuracy.

Alternatively, in the fixed-guided PVDF sensor as depicted in Fig. 2.7d, PVDF films are used to replace the rigid flexure in multi-stage compound parallelogram flexure (MCPF) to measure the pure one-dimensional force and hold the cell stably. However, the resolution is relatively low in several hundred μN [101].

In addition, the manual assembly in the aforementioned force sensors can induce misalignments and produce significant errors in force measurement [35].

2.6 Current Challenges on Cell Microinjection

In order to achieve automated cell microinjection manipulation, there are still many challenges in other respects besides the force sensor. The trends in many micromanipulation applications require that the future micromanipulation systems should address the following challenges. The whole cell injection system should satisfy the requirements of high speed, high flexibility, high level of automation, large information content, as well as low cost. Aiming at improving the success rate of cell injection and survive rate of injected cells, some challenges needed to be overcome and some promising methods are discussed as follows.

2.6.1 Micromanipulator Design

The micromanipulator is expected to implement the planned task with a high accuracy and repeatability in a short time without damaging the cells. To realize the automation process, the maximum speed of the micromanipulator should be a few millimeters per second. The forces applied by the micromanipulator should be in μN level. The micromanipulator should has the ability to complete the manipulation tasks without deteriorating the motion control performance. The aforementioned performance requirements should be considered in both hardware and software design. Specifically, the selection of actuators and sensors and design of mechanical structure should be conducted by taking into account the performance requirements. In addition, the motion control software and related algorithms must be able to achieve the desired performance and possibly compensate for the defects (e.g., nonlinearity and uncertainty) of the actuators. The micromanipulator must be controlled to enable both transient and steady-state requirements in the position and contact force response for the end-effectors.

Generally, the positioning challenges in automatic micromanipulation can be solved by adopting suitable sensors and actuators that offer sufficient resolution, speed, stroke, and compact size [28]. Moreover, cooperative control of two micromanipulators should be performed to maximize the operation speed [99]. As a frequently used actuator in micromanipulator, piezoelectric actuator provides rapid response and ultrahigh motion resolution, whereas it also introduces nonlinearities dominated by hysteresis and drift effects. Sophisticated control algorithms are required to realize a precise motion/force output [111].

2.6.2 Injection Control Design

In the injection manipulation, the dimension and location of the cell to be injected should be determined through computer vision technique first. Then, the needle is moved to approach the cell, which is controlled by visual servo control. In the force control scheme, when the needle begins to contact the cell, the force control will start. The switch between the position and force control should be smooth during the injection [109]. The injection control should be robust enough to tolerate various disturbances. For instance, there is certain stiffness variance for a batch of cells in the same living stage and the stiffness of a living cell changes at different stages of life [32].

During the cell injection process, the force control mode is used to avoid damaging the cells, while position control mode is adopted when there is no force exerted. The position/force switching control should be carefully designed to avoid adverse effects such as oscillation and overshoot, which will damage the cells. The impedance control has emerged as a promising method to perform a smooth switch between position and force control for micromanipulation [21]. The impedance control can control the free motion and contact force by using a single impedance algorithm [66]. Moreover, a weight coefficient method has been presented in [98] to smooth the switch process between position and force control.

2.6.3 Cell Holder Design

Currently, there are many cell trapping techniques using surface chemistry [120], dielectrophoresis, optical tweezers, ultrasonic trapping, magnetic trapping, and mechanical confinements to trap and immobilize cells [48]. Among these techniques, only the mechanical confinements are able to provide adequate immobilization forces for microinjection applications. These mechanical confinement structures include microwells, hydrodynamic traps, and vacuum-based confinements [48].

The cells trapped in microwells can tinily move inside the microwells, resulting in unsecured immobilization during the cell injection. Cells trapped in the hydrodynamic microchannels prevent the injection micropipette from accessing cells. Thus, the vacuum-based confinements are widely used in cell injection. Traditionally, a holding micropipette is applied to locate randomly dispersed cells in the petri dish and transport the cell to the injection area once at a time, which limits the throughput and efficiency. Hence, the development of a cell holder that is able to quickly fix multiple cells into a regular pattern can improve the efficiency of both manual and robotic cell microinjection operations. In practice, differential interference contrast (DIC) microscope is most commonly utilized in microinjection for cell imaging, and glass is the most preferred material to fabricate cell immobilization devices [64].

In the literature, Fujisato [16] has proposed a special cell holder. It is made of microporous glass (MPG) with a glass or stainless coating layer, on which sand-blasted micropocket holes are fabricated. Then, liquid containing cells is pumped into the cell holder and the cells stay in the holes with liquid flow through the microporous glass, as shown in the Fig. 2.8a. Lu et al. [52] have presented a gel-based cell holder with several parallel V-shaped grooves to fix cells, as shown in the Fig. 2.8b. However, the cells can easily move if the force direction is not purely perpendicular to the plane of the cell holder. Huang et al. [23] have designed a circular and rotary plate, mainly composed of hemispherical hole and groove to fix many cells by mechanical confinements and maintain the cells wet during cell injection, as shown in the Fig. 2.8c. However, the cells need to be located in the hemispherical hole manually,

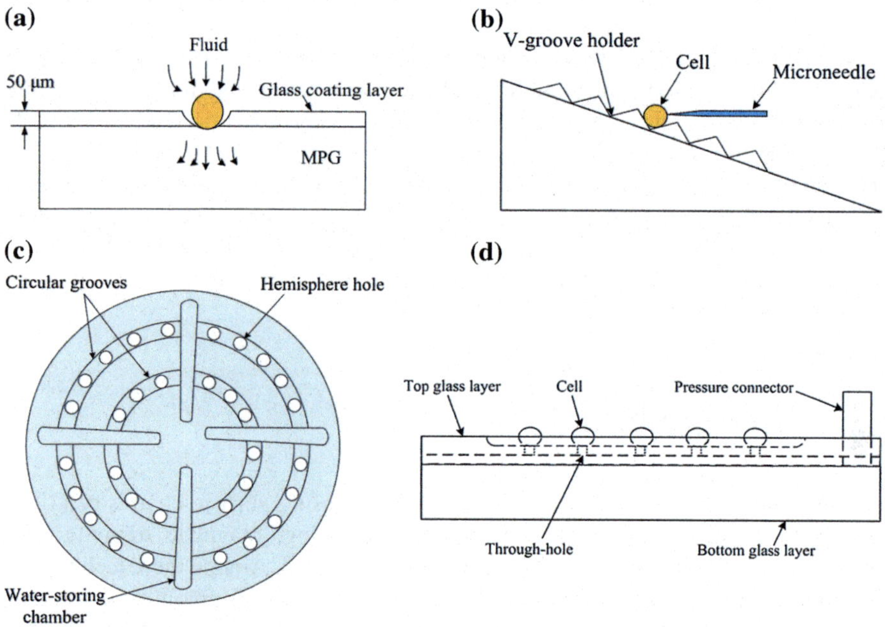

Fig. 2.8 Typical structures of mechanical confinement cell holders. **a** Microporous glass (MSG) with holes as cell holder, **b** parallel V-groove based gel cell holder, **c** hemispherical hole and groove based circular rotary cell holder, **d** vacuum-based through-hole cell holder

which increases the risk of cell damage and decreases the efficiency inherently. Liu et al. [48] have reported a glass-based cell holding device for single cell fixation by evenly spaced through-holes linked with a low vacuum source, as shown in Fig. 2.8d.

In summary, the cell holder should be designed to fix many cells quickly and stably without increasing the risk of damage to the cells.

2.6.4 Penetration Scheme Design

Usually, the cells will become inactivate when the deformation of the injected cells exceeds a certain limit. If the micropipette injects the cell at a constant speed, the cell may experience an extra large deformation until the membrane is penetrated. Usually, an impact actuator [91, 112], which is able to produce an accelerated and fast movement, can be used to reduce the deformation. Moreover, the frictionless compound flexure stage [51] can also be used to directly control the injection force to produce a variable speed and fast acceleration [52]. Consequently, the cell membrane should be penetrated with a high speed with acceleration and rapid deceleration after the membrane is penetrated. However, the process of rapid acceleration and deceleration can cause vibration, which should be eliminated in practice. Although the requirements on high acceleration/deceleration and the elimination of vibrations collide with each other, they should be compromised to achieve a successful injection. In the literature, Huang et al. [21] have presented a velocity and acceleration profiles of the injecting pipette during the injection to inject zebrafish embryos.

Moreover, in order to improve the survival rate of the injected cell, the deformation of the cell during the penetration should be reduced as much as possible. The penetration of the cell membrane can be produced by either linear or vibration motion of the microinjecting pipette. Minimization of the penetration force is vital to the development of high-throughput micromanipulation facilities for biology and genetics research, such as RNAi for gene silencing. Vibration is a popular method for decreasing the cutting force of macroscopic tools. Penetration, produced by either increased vibration (high-frequency small amplitude vibration) or translation, dramatically reduces the force on the injector [118]. It has been shown that the combination of vibration and linear translation greatly reduces the injection force when penetrating Drosophila embryos [118].

It is notable that piezoelectric actuators are very popular in cell injection manipulation thanks to their smart structure and high control bandwidth [91]. A piezo-driven pipette has been utilized to implement ICSI in mouse [36], which shows a survival rate of 80% for sperm-injected oocytes. In addition, piezodrill-based injection has been proved as an effective approach for nuclear transfer between horse and cattle oocytes [12].

2.6.5 Injecting Pipette Maintenance

The injecting pipettes can be fabricated out of tiny glass tube using a commercial pipette puller. The drawn glass pipettes have large and undefined needle diameters, because the drawn pipette is closed and must be broke off at desired location for injection use. Such a process is usually done manually. Moreover, the fracture surface is often greatly rough, leading to severe damage to injected cells. Furthermore, glass injecting pipettes are long and mechanically fragile, which produces high back-pressures [115] and degrades the repeatability of injection [115]. In the literature, Li et al. [42] have proposed a probe-etching technique to control the diameter of fiber probes, ranging from 500 nm to 1.5 μm. Zhang et al. [120] have demonstrated the high injection efficiency of the silicon nitride injector as compared with traditional glass-drawn needle.

As the injecting pipette is easy to be broken and get clogged, the condition of pipette should be monitored to improve the reliability [28]. In current cell injection systems, the user must clean the injecting pipette after each injection by using an abundant pressure pulse and manually check if a liquid jet comes out from the injecting pipette. Alternatively, the injecting pipette can be changed constantly without checking, which is very labor consuming. Furthermore, the breakage of injecting pipette tip is frequently occurred. Consequently, a method to monitor the condition of injecting pipette is necessary. Possible techniques to solve the problem involves machine vision and electrical method. For example, Lukkari et al. [53] have proposed an impedance measurement method to detect the condition of break, clogging, and faulty injection solution, which is based on electrical method.

Moreover, since the injecting pipette can become blunted or broken frequently during the cell injection, it should be replaced with a new one. However, this work is also burdensome and deteriorates the throughout and efficiency. Thus, a fast exchanger for the injecting pipette is necessary. For instance, Matsuoka et al. [56] suggested the preparation of a number of injecting pipettes with the same length in advance for replacement once the injecting pipette is broken. It should also be mentioned that the sensor should be separated from the injecting pipette to allow a quick change of the broken pipette.

2.6.6 Injection Volume Issue

Ideally, if the injection pressure, injection time, and balance pressure keep the same value, all of the injected cells can be injected with the same volume of the injection materials [28]. Nevertheless, in fact, the volume varies significantly [61]. The efflux from or influx into the injecting pipette also affects the repeatability of injected volume of liquid. In the case of influx, the cell medium flows into the injecting pipette, leading to the injection of less than desired volume of solution. In the case of efflux, the injection solution leaks out from the injecting pipette, causing more volume transferred than desired value. The variation caused by the influx and efflux may lead to the unexpected change of experimental results. Furthermore, influx can also give rise to the clogging of the injecting pipette by aggregating in the culture solution, and hence changes the opening size of the injecting pipette or even completely prevents the injection process. In order to avoid injecting pipette clogging, most researchers have adopted the efflux of the filling solution rather than the influx of the culture medium [28].

Some researchers also adopt calibration methods, such as oil droplet method, to guarantee the injection volume [109]. Nevertheless, the model in the calibration is influenced by the equipment-related disturbance parameters and biological disturbance parameters. The relationship between the injection parameters and the disturbance parameters has been analyzed in [38]. Moreover, Zhang et al. [118] have directly integrated a microfabricated injecting pipette with a piezoresistive pressure sensor to control of the injection volume.

Hence, an oil droplet method is a promising method to calibrate the injecting volume for practical applications. In addition, experimental and theoretical results have indicated that the efflux at the entrance of the injecting pipette is the main reason of acute cell death. Alginate and nanodiamonds are two promising carriers to reduce the efflux after the mechanically injection operation using the injecting pipette. These methods have the potential to improve the survival rate and transplantation rate of the injected cells [1].

2.7 Conclusion

Cell injection plays an important role in genetics, transgenics, molecular biology, drug discovery, reproductive studies, and other biomedical fields. Robotic cell injection with force feedback can improve the success rate and survival rate of cell injection, as compared with manual cell injection and robotic cell injection with position feedback only. Thus, the development of high-performance force sensors has been a hot topic in recent years. In this chapter, the most frequently used force sensors in robotic cell injection have been extensively reviewed to address the research status toward the challenges. Firstly, the role of cell injection, limitation of manual cell injection, and current methods of cell injection are reviewed in detail. Then, the commercial and research development of adherent cell injection are introduced, and the commonly used suspended cell models are presented detailedly. The significance of force sensors in robotic cell injection system is also presented. Next, five most popular force-sensing methods in cell microinjection are reviewed. Finally, some remaining challenges and some promising solutions are discussed.

References

1. Aguado, B.A., Mulyasasmita, W., Su, J., Lampe, K.J., Heilshorn, S.C.: Improving viability of stem cells during syringe needle flow through the design of hydrogel cell carriers. Tissue Eng. Part A **18**(7–8), 806–815 (2011)
2. Ammi, M., Ferreira, A.: Realistic visual and haptic rendering for biological-cell injection. In: Proceedings of the 2005 IEEE International Conference on Robotics and Automation (ICRA 2005), pp. 918–923 (2005)
3. Ammi, M., Ferreira, A.: Biological cell injection visual and haptic interface. Adv. Robot. **20**(3), 283–304 (2006)
4. Ammi, M., Ladjal, H., Ferreira, A.: Evaluation of 3D pseudo-haptic rendering using vision for cell micromanipulation. In: Proceedings of the 2006 IEEE/RSJ International Conference on Intelligent Robots and Systems, pp. 2115–2120 (2006)
5. Asgari, M., Abdi, H., Lim, C.P., Nahavandi, S.: Formulation and simulation of a 3D mechanical model of embryos for microinjection. In: Proceedings of the 2013 IEEE International Conference on Systems, Man, and Cybernetics (SMC), pp. 2219–2224 (2013)
6. Banga, A.K., Bose, S., Ghosh, T.K.: Iontophoresis and electroporation: comparisons and contrasts. Int. J. Pharm. **179**(1), 1–19 (1999)
7. Beutel, T., Ferreira, N., Balck, A., Leester-Schadel, M., Buttgenbach, S.: Cell manipulation system based on a self-calibrating silicon micro force sensor providing capillary status monitoring. IEEE Sens. J. **12**(10), 3075–3081 (2012)
8. Beyeler, F., Muntwyler, S., Nelson, B.J.: A six-axis mems force-torque sensor with micro-newton and nano-newtonmeter resolution. J. Microelectromech. Syst. **18**(2), 433–441 (2009)
9. Canny, J.: A computational approach to edge detection. IEEE Trans. Pattern Anal. Mach. Intell. **6**, 679–698 (1986)
10. Carthew, R.W.: Gene silencing by double-stranded RNA. Curr. Opin. Cell Biol. **13**(2), 244–248 (2001)
11. Chen, Z., Shen, Y., Xi, N., Tan, X.: Integrated sensing for ionic polymer-metal composite actuators using PVDF thin films. Smart Mater. Struct. **16**(2), S262 (2007)
12. Choi, Y., Love, C., Chung, Y., Varner, D., Westhusin, M., Burghardt, R., Hinrichs, K.: Production of nuclear transfer horse embryos by Piezo-driven injection of somatic cell nuclei and activation with stallion sperm cytosolic extract. Biol. Reprod. **67**(2), 561–567 (2002)
13. Davis, L.S.: A survey of edge detection techniques. Comput. Graph. Image Process. **4**(3), 248–270 (1975)
14. Desmaele, D., Boukallel, M., Regnier, S.: A planar structure sensitive to out-of-plane forces for the force-controlled injection of suspended and adherent cells. In: Proceedings of the 2011 Annual International Conference of the IEEE Engineering in Medicine and Biology Society (EMBC), pp. 8420–8423 (2011)
15. Fauver, M.E., Dunaway, D.L., Lilienfeld, D.H., Craighead, H.G., Pollack, G.H.: Microfabricated cantilevers for measurement of subcellular and molecular forces. IEEE Trans. Biomed. Eng. **45**(7), 891–898 (1998)
16. Fujisato, T., Abe, S., Tsuji, T., Sada, M., Miyawaki, F., Ohba, K.: The development of an OVA holding device made of microporous glass plate for genetic engineering. In: Proceedings of the 20th Annual International Conference of the IEEE Engineering in Medicine and Biology Society, vol. 6, pp. 2981–2982 (1998)
17. Fung, C.K., Elhajj, I., Li, W.J., Xi, N.: A 2-D PVDF force sensing system for micro-manipulation and micro-assembly. In: Proceedings of IEEE International Conference on Robotics and Automation (ICRA'02), vol. 2, pp. 1489–1494 (2002)
18. Gauthier, M., Nourine, M.: Capillary force disturbances on a partially submerged cylindrical micromanipulator. IEEE Trans. Rob. **23**(3), 600–604 (2007)
19. Ghanbari, A., Horan, B., Nahavandi, S., Chen, X., Wang, W.: Haptic microrobotic cell injection system. IEEE Syst. J. **8**(2), 371–383 (2014)
20. Greminger, M.A., Nelson, B.J.: Vision-based force measurement. IEEE Trans. Pattern Anal. Mach. Intell. **26**(3), 290–298 (2004)

21. Huang, H., Sun, D., Mills, J.K., Li, W.J., Cheng, S.H.: Visual-based impedance control of out-of-plane cell injection systems. IEEE Trans. Autom. Sci. Eng. **6**(3), 565–571 (2009)
22. Huang, H., Sun, D., Su, H., Mills, J.K.: Force sensing and control in robot-assisted suspended cell injection system. Adv. Rob. Virtual Reality 61–88 (2012)
23. Huang, H.B., Sun, D., Mills, J.K., Cheng, S.H.: Robotic cell injection system with position and force control: toward automatic batch biomanipulation. IEEE Trans. Rob. **25**(3), 727–737 (2009)
24. Huang, T., Kimura, Y., Yanagimachi, R.: The use of piezo micromanipulation for intracytoplasmic sperm injection of human oocytes. J. Assist. Reprod. Genet. **13**(4), 320–328 (1996)
25. Ingber, D.E.: Tensegrity: the architectural basis of cellular mechanotransduction. Annu. Rev. Physiol. **59**(1), 575–599 (1997)
26. Juelicher, F., Kruse, K., Prost, J., Joanny, J.F.: Active behavior of the cytoskeleton. Phys. Rep. **449**(1), 3–28 (2007)
27. Jumpertz, R., Hart, A., Ohlsson, O., Saurenbach, F., Schelten, J.: Piezoresistive sensors on AFM cantilevers with atomic resolution. Microelectron. Eng. **41**, 441–444 (1998)
28. Kallio, P., Kuncova-Kallio, J.: Capillary pressure microinjection of living adherent cells: challenges in automation. J. Micromechatronics **3**(3), 189–220 (2006)
29. Karimirad, F., Chauhan, S., Shirinzadeh, B.: Vision-based force measurement using neural networks for biological cell microinjection. J. Biomech. **47**(5), 1157–1163 (2014)
30. Kass, M., Witkin, A., Terzopoulos, D.: Snakes: active contour models. Int. J. Comput. Vis. **1**(4), 321–331 (1988)
31. Kim, D.H., Kim, B., Yun, S., Kwon, S.: Cellular force measurement for force reflected biomanipulation. In: Proceedings of the 2004 IEEE International Conference on Robotics and Automation (ICRA'04), vol. 3, pp. 2412–2417 (2004)
32. Kim, D.H., Sun, Y., Yun, S., Kim, B., Hwang, C.N., Lee, S.H., Nelson, B.J.: Mechanical property characterization of the zebrafish embryo chorion. In: Proceedings of the 26th Annual International Conference of the IEEE Engineering in Medicine and Biology Society (IEMBS'04), vol. 2, pp. 5061–5064 (2004)
33. Kim, J., Janabi-Sharifi, F., Kim, J.: Haptic feedback based on physically based modeling for cellular manipulation systems. In: Proceedings of the International Conference on Human Haptic Sensing and Touch Enabled Computer Applications, pp. 661–667 (2008)
34. Kim, J., Janabi-Sharifi, F., Kim, J.: A haptic interaction method using visual information and physically based modeling. IEEE/ASME Trans. Mechatron. **15**(4), 636–645 (2010)
35. Kim, K., Liu, X., Zhang, Y., Sun, Y.: Nanonewton force-controlled manipulation of biological cells using a monolithic MEMS microgripper with two-axis force feedback. J. Micromech. Microeng. **18**(5), 055013 (2008)
36. Kimura, Y., Yanagimachi, R.: Intracytoplasmic sperm injection in the mouse. Biol. Reprod. **52**(4), 709–720 (1995)
37. Kuncova, J., Kallio, P.: Challenges in capillary pressure microinjection. In: Proceedings of the 26th Annual International Conference of the IEEE Engineering in Medicine and Biology Society (IEMBS'04), vol. 2, pp. 4998–5001 (2004)
38. Kuncova-Kallio, J.: Calibration of a capillary pressure microinjection and its implementation within a micromanipulator. M.Sc. thesis, Tampere University of Technology (2002)
39. Ladjal, H., Hanus, J.L., Ferreira, A.: Micro-to-nano biomechanical modeling for assisted biological cell injection. IEEE Trans. Biomed. Eng. **60**(9), 2461–2471 (2013)
40. Langenau, D.M., Keefe, M.D., Storer, N.Y., Guyon, J.R., Kutok, J.L., Le, X., Goessling, W., Neuberg, D.S., Kunkel, L.M., Zon, L.I.: Effects of RAS on the genesis of embryonal rhabdomyosarcoma. Genes Dev. **21**(11), 1382–1395 (2007)
41. Leymarie, F., Levine, M.D.: Tracking deformable objects in the plane using an active contour model. IEEE Trans. Pattern Anal. Mach. Intell. **15**(6), 617–634 (1993)
42. Li, W.J., Xi, N.: Novel micro gripping, probing, and sensing devices for single-cell surgery. In: Proceedings of the 26th Annual International Conference of the IEEE Engineering in Medicine and Biology Society (IEMBS'04), vol. 1, pp. 2591–2594 (2004)

43. Li, X., Zong, G., Bi, S.: Development of global vision system for biological automatic micromanipulation system. In: Proceedings of the 2001 IEEE International Conference on Robotics and Automation (ICRA), vol. 1, pp. 127–132 (2001)
44. Li, X., Zong, G., Bi, S., Zhao, W.: Automatic micromanipulating system for biological applications with visual servo control. J. Micromechatron. **1**(4), 345–363 (2001)
45. Lim, C., Zhou, E., Quek, S.: Mechanical models for living cells—a review. J. Biomech. **39**(2), 195–216 (2006)
46. Lin, M.T., Pulkkinen, L., Uitto, J., Yoon, K.: The gene gun: current applications in cutaneous gene therapy. Int. J. Dermatol. **39**(3), 161–170 (2000)
47. Liu, X., Kim, K., Zhang, Y., Sun, Y.: Nanonewton force sensing and control in microrobotic cell manipulation. Int. J. Rob. Res. **28**(8), 1065–1076 (2009)
48. Liu, X., Sun, Y.: Microfabricated glass devices for rapid single cell immobilization in mouse zygote microinjection. Biomed. Microdevices **11**(6), 1169 (2009)
49. Liu, X., Sun, Y., Wang, W., Lansdorp, B.M.: Vision-based cellular force measurement using an elastic microfabricated device. J. Micromech. Microeng. **17**(7), 1281 (2007)
50. Loh, O., Lam, R., Chen, M., Moldovan, N., Huang, H., Ho, D., Espinosa, H.D.: Nanofountain-probe-based high-resolution patterning and single-cell injection of functionalized nanodiamonds. Small **5**(14), 1667–1674 (2009)
51. Lu, Z., Chen, P.C., Lin, W.: Force sensing and control in micromanipulation. IEEE Trans. Syst. Man Cybern. Part C Appl. Rev. **36**(6), 713–724 (2006)
52. Lu, Z., Chen, P.C., Nam, J., Ge, R., Lin, W.: A micromanipulation system with dynamic force-feedback for automatic batch microinjection. J. Micromech. Microeng. **17**(2), 314 (2007)
53. Lukkari, M., Kallio, P.: Multi-purpose impedance-based measurement system to automate microinjection of adherent cells. In: Proceedings of the 2005 IEEE International Symposium on Computational Intelligence in Robotics and Automatio (CIRA 2005), pp. 701–706 (2005)
54. Marr, D., Hildreth, E.: Theory of edge detection. Proc. R. Soc. Lond. B Biol. Sci. **207**(1167), 187–217 (1980)
55. Maruyama, H., Masuda, T., Liu, H., Arai, F.: Selective and rapid cell injection of fluorescence sensor encapsulated in liposome using optical control of zeta potential and local mechanical stimulus by optical tweezers. In: Proceedings of the 2014 IEEE/RSJ International Conference on Intelligent Robots and Systems (IROS 2014), pp. 816–821 (2014)
56. Matsuoka, H., Komazaki, T., Mukai, Y., Shibusawa, M., Akane, H., Chaki, A., Uetake, N., Saito, M.: High throughput easy microinjection with a single-cell manipulation supporting robot. J. Biotechnol. **116**(2), 185–194 (2005)
57. Matsuoka, H., Shimoda, S., Miwa, Y., Saito, M.: Automatic positioning of a microinjector in mouse es cells and rice protoplasts. Bioelectrochemistry **69**(2), 187–192 (2006)
58. Mehrbod, M., Mofrad, M.R.: On the significance of microtubule flexural behavior in cytoskeletal mechanics. PloS One **6**(10), e25627 (2011)
59. Meister, A., Gabi, M., Behr, P., Studer, P., Voros, J., Niedermann, P., Bitterli, J., Polesel-Maris, J., Liley, M., Heinzelmann, H., et al.: FluidFM: combining atomic force microscopy and nanofluidics in a universal liquid delivery system for single cell applications and beyond. Nano Lett. **9**(6), 2501–2507 (2009)
60. Meyer, G., Amer, N.M.: Novel optical approach to atomic force microscopy. Appl. Phys. Lett. **53**(12), 1045–1047 (1988)
61. Minaschek, G., Bereiter-Hahn, J., Bertholdt, G.: Quantitation of the volume of liquid injected into cells by means of pressure. Exp. Cell Res. **183**(2), 434–442 (1989)
62. Muntwyler, S., Beyeler, F., Nelson, B.: Three-axis micro-force sensor with sub-micro-Newton measurement uncertainty and tunable force range. J. Micromech. Microeng. **20**(2), 025011 (2009)
63. Muraoka, O., Shimizu, T., Yabe, T., Nojima, H., Bae, Y.K., Hashimoto, H., Hibi, M.: Sizzled controls dorso-ventral polarity by repressing cleavage of the Chordin protein. Nat. Cell Biol. **8**(4), 329 (2006)
64. Murphy, D.B.: Fundamentals of light microscopy and electronic imaging. Wiley, New York (2002)

65. Nakayama, T., Fujiwara, H., Tastumi, K., Fujita, K., Higuchi, T., Mori, T.: A new assisted hatching technique using a piezo-micromanipulator. Fertil. Steril. **69**(4), 784–788 (1998)
66. Part, S.: Impedance control: an approach to manipulation. J. Dyn. Syst. Meas. Contr. **107**, 17 (1985)
67. Permana, S., Grant, E., Walker, G.M., Yoder, J.A.: A review of automated microinjection systems for single cells in the embryogenesis stage. IEEE/ASME Trans. Mechatron. **21**(5), 2391–2404 (2016)
68. Pillarisetti, A., Anjum, W., Desai, J.P., Friedman, G., Brooks, A.D.: Force feedback interface for cell injection. In: Proceedings of the First Joint Eurohaptics Conference and Symposium on Haptic Interfaces for Virtual Environment and Teleoperator Systems, pp. 391–400 (2005)
69. Pillarisetti, A., Pekarev, M., Brooks, A.D., Desai, J.P.: Evaluating the effect of force feedback in cell injection. IEEE Trans. Autom. Sci. Eng. **4**(3), 322 (2007)
70. Poleo, G.A., Denniston, R.S., Reggio, B.C., Godke, R.A., Tiersch, T.R.: Fertilization of eggs of zebrafish, *Danio rerio*, by intracytoplasmic sperm injection. Biol. Reprod. **65**(3), 961–966 (2001)
71. Rols, M.P.: Electropermeabilization, a physical method for the delivery of therapeutic molecules into cells. Biochim. Biophys. Acta (BBA) Biomembr. **1758**(3), 423–428 (2006)
72. Rubin, G.M., Lewis, E.B.: A brief history of drosophila's contributions to genome research. Science **287**(5461), 2216–2218 (2000)
73. Rubin, G.M., Yandell, M.D., Wortman, J.R., Gabor, G.L., Nelson, C.R., Hariharan, I.K., Fortini, M.E., Li, P.W., Apweiler, R., Fleischmann, W., et al.: Comparative genomics of the eukaryotes. Science **287**(5461), 2204–2215 (2000)
74. Saffrich, R., Pepperkok, R.: Embo practical course–microinjection and detection of probes in cells (2001). http://www.bio.net/mm/cellbiol/1998-March/008361.html
75. Scherp, P., Hasenstein, K.: Microinjection—a tool to study gravitropism. Adv. Space Res. **31**(10), 2221–2227 (2003)
76. Shen, Y., Wejinya, U., Xi, N., Pomeroy, C.A.: Force measurement and mechanical characterization of living drosophila embryos for human medical study. Proc. Inst. Mech. Eng. Part H J. Eng. Med. **221**(2), 99–112 (2007)
77. Shen, Y., Winder, E., Xi, N., Pomeroy, C.A., Wejinya, U.C.: Closed-loop optimal control-enabled piezoelectric microforce sensors. IEEE/ASME Trans. Mechatron. **11**(4), 420–427 (2006)
78. Shim, J.H., Cho, S.Y., Cha, D.H.: Vision-guided micromanipulation system for biomedical application. Proc. SPIE **5604**, 98–107 (2004)
79. Shulev, A., Roussev, I., Kostadinov, K.: Force sensor for cell injection and characterization. In: Proceedings of the 2012 International Conference on Manipulation, Manufacturing and Measurement on the Nanoscale (3M-NANO), pp. 335–338 (2012)
80. Stavrov, V.T., Shulev, A.A., Hardalov, C.M., Todorov, V.M., Roussev, I.R.: All-silicon microforce sensor for bio applications. In: Proceedings of SPIE, vol. 8763, p. 87630Y (2013)
81. Sun, Y., Fry, S.N., Potasek, D., Bell, D.J., Nelson, B.J.: Characterizing fruit fly flight behavior using a microforce sensor with a new comb-drive configuration. J. Microelectromech. Syst. **14**(1), 4–11 (2005)
82. Sun, Y., Nelson, B.J.: Biological cell injection using an autonomous microrobotic system. Int. J. Rob. Res. **21**(10–11), 861–868 (2002)
83. Sun, Y., Nelson, B.J.: Mems for cellular force measurements and molecular detection. Int. J. Inf. Acquisition **1**(01), 23–32 (2004)
84. Sun, Y., Nelson, B.J.: Mems capacitive force sensors for cellular and flight biomechanics. Biomed. Mater. **2**(1), S16 (2007)
85. Sun, Y., Nelson, B.J., Potasek, D.P., Enikov, E.: A bulk microfabricated multi-axis capacitive cellular force sensor using transverse comb drives. J. Micromech. Microeng. **12**(6), 832 (2002)
86. Sun, Y., Wan, K.T., Roberts, K.P., Bischof, J.C., Nelson, B.J.: Mechanical property characterization of mouse zona pellucida. IEEE Trans. Nanobiosci. **2**(4), 279–286 (2003)
87. Sun, Z., Hao, L., Chen, W., Li, Z.: Robotic cell injection force control based on static PVDF sensor and fuzzy-PID control method. Int. J. Appl. Electromagnet. Mech. **41**(1), 73–86 (2013)

88. Sundaram, J., Mellein, B.R., Mitragotri, S.: An experimental and theoretical analysis of ultrasound-induced permeabilization of cell membranes. Biophys. J. **84**(5), 3087–3101 (2003)
89. Tan, Y., Sun, D., Huang, W., Cheng, S.H.: Mechanical modeling of biological cells in microinjection. IEEE Trans. Nanobiosci. **7**(4), 257–266 (2008)
90. Tan, Y., Sun, D., Huang, W., Cheng, S.H.: Characterizing mechanical properties of biological cells by microinjection. IEEE Trans. Nanobiosci. **9**(3), 171–180 (2010)
91. Tanikawa, T., Arai, T.: Development of a micro-manipulation system having a two-fingered micro-hand. IEEE Trans. Rob. Autom. **15**(1), 152–162 (1999)
92. Tortonese, M., Yamada, H., Barrett, R., Quate, C.: Atomic force microscopy using a piezoresistive cantilever. In: Proceedings of the 1991 International Conference on Solid-State Sensors and Actuators (TRANSDUCERS'91), pp. 448–451 (1991)
93. Tran, N.D., Liu, X., Yan, Z., Abbote, D., Jiang, Q., Kmiec, E.B., Sigmund, C.D., Engelhardt, J.F.: Efficiency of chimeraplast gene targeting by direct nuclear injection using a GFP recovery assay. Mol. Ther. **7**(2), 248–253 (2003)
94. Ujihara, Y., Nakamura, M., Miyazaki, H., Wada, S.: Proposed spring network cell model based on a minimum energy concept. Ann. Biomed. Eng. **38**(4), 1530–1538 (2010)
95. Wall, R.J.: Pronuclear microinjection. Cloning Stem Cells **3**(4), 209–220 (2001)
96. Walter, W., Stein, U.: Viral vectors for gene transfer a review of their use in the treatment of human disease. Drugs **60**, 249–71 (2000)
97. Wan, H.I., DiAntonio, A., Fetter, R.D., Bergstrom, K., Strauss, R., Goodman, C.S.: Highwire regulates synaptic growth in *Drosophila*. Neuron **26**(2), 313–329 (2000)
98. Wang, G., Xu, Q.: Design and precision position/force control of a piezo-driven microinjection system. IEEE/ASME Trans. Mechatron. **22**(4), 1744–1754 (2017)
99. Wang, W., Liu, X., Gelinas, D., Ciruna, B., Sun, Y.: A fully automated robotic system for microinjection of zebrafish embryos. PloS One **2**(9), e862 (2007)
100. Wei, Y., Xu, Q.: An overview of micro-force sensing techniques. Sens. Actuators A Phys. **234**, 359–374 (2015)
101. Wei, Y., Xu, Q.: Design of a PVDF-MFC force sensor for robot-assisted single cell microinjection. IEEE Sens. J. **17**(13), 3975–3982 (2017)
102. Wejinya, U.C., Shen, Y., Xi, N., Salem, F.: Force measurement of embryonic system using in situ PVDF piezoelectric sensor. In: Proceedings of the 49th IEEE International Midwest Symposium on Circuits and Systems (MWSCAS'06), vol. 1, pp. 108–112 (2006)
103. Wenming, X., Hui, Z.: Bio-manipulation probe integration with micro-force sensor. In: Proceedings of the 3rd IEEE International Conference on Nano/Micro Engineered and Molecular Systems (NEMS 2008), pp. 393–396 (2008)
104. Westerfield, M.: The zebrafish book: a guide for the laboratory use of zebrafish. http://zfin.org/zf_info/zfbook/zfbk.html (2000)
105. Wiens, G.J., Roman, G.A.: A mechanism approach for enhancing the dynamic range and linearity of MEMS optical force sensing. Signal Measurement and Estimation Techniques for Micro and Nanotechnology, p. 193 (2011)
106. Wright, W.H., Sonek, G., Tadir, Y., Berns, M.W.: Laser trapping in cell biology. IEEE J. Quant. Electron. **26**(12), 2148–2157 (1990)
107. Xie, Y., Sun, D., Liu, C.: Penetration force measurement and control in robotic cell microinjection. In: Proceedings of the 2009 IEEE/RSJ International Conference on Intelligent Robots and Systems, pp. 4701–4706 (2009)
108. Xie, Y., Sun, D., Liu, C., Tse, H.Y., Cheng, S.H.: A force control approach to a robot-assisted cell microinjection system. Int. J. Rob. Res. **29**(9), 1222–1232 (2010)
109. Xie, Y., Sun, D., Tse, H.Y.G., Liu, C., Cheng, S.H.: Force sensing and manipulation strategy in robot-assisted microinjection on zebrafish embryos. IEEE/ASME Trans. Mechatron. **16**(6), 1002–1010 (2011)
110. Xu, Q.: Design and development of a novel compliant gripper with integrated position and grasping/interaction force sensing. IEEE Trans. Autom. Sci. Eng. (2015)
111. Xu, Q., Tan, K.K.: Advanced Control of Piezoelectric Micro-/Nano-Positioning Systems. Springer, Berlin (2015)

112. Yamagata, Y., Higuchi, T.: A micropositioning device for precision automatic assembly using impact force of piezoelectric elements. In: Proceedings of the 1995 IEEE International Conference on Robotics and Automation, vol. 1, pp. 666–671 (1995)
113. Yu, S., Nelson, B.J.: Microrobotic cell injection. In: Proceedings of the 2001 IEEE International Conference on Robotics and Automation (ICRA), vol. 1, pp. 620–625 (2001)
114. Yunlei, Z., Shuang, L., Chongjun, Y., Chong, L., Yu, X.: Design of a microforce sensor based on fixed-simply supported beam: towards realtime cell microinjection. In: Proceeedings of the 2015 IEEE International Conference on Cyber Technology in Automation, Control, and Intelligent Systems (CYBER), pp. 1080–1084 (2015)
115. Zappe, S., Fish, M., Scott, M.P., Solgaard, O.: Automated mems-based drosophila embryo injection system for high-throughput rnai screens. Lab Chip **6**(8), 1012–1019 (2006)
116. Zeng, Y., Yip, A.K., Teo, S.K., Chiam, K.H.: A three-dimensional random network model of the cytoskeleton and its role in mechanotransduction and nucleus deformation. Biomech. Model. Mechanobiol. **11**(1–2), 49–59 (2012)
117. Zhang, W., Sobolevski, A., Li, B., Rao, Y., Liu, X.: An automated force-controlled robotic micromanipulation system for mechanotransduction studies of drosophila larvae. IEEE Trans. Autom. Sci. Eng. **13**(2), 789–797 (2016)
118. Zhang, X., Scott, M.P., Quate, C.F., Solgaard, O.: Microoptical characterization of piezoelectric vibratory microinjections in drosophila embryos for genome-wide RNAi screen. J. Microelectromech. Syst. **15**(2), 277–286 (2006)
119. Zhang, X., Zappe, S., Bernstein, R., Sahin, O., Chen, C.C., Fish, M., Scott, M., Solgaard, O.: Integrated optical diffractive micrograting-based injection force sensor. In: Proceedings of the 12th International Conference on Solid-State Sensors, Actuators and Microsystems (TRANSDUCERS), vol. 2, pp. 1051–1054 (2003)
120. Zhang, X., Zappe, S., Bernstein, R., Sahin, O., Chen, C.C., Fish, M., Scott, M., Solgaard, O.: Micromachined silicon force sensor based on diffractive optical encoders for characterization of microinjection. Sens. Actuators A Phys. **114**(2), 197–203 (2004)

Chapter 3
Design, Fabrication, and Testing of a Microforce Sensor for Microinjection

Abstract In this chapter, a novel one-degree-of-freedom (1-DOF) microforce sensor is designed by adopting macrofiber composite (MFC) and polyvinylidene fluoride (PVDF) films. The films function as fixed-guided beams and construct a multistage compound parallelogram flexure mechanism. The sensor enables the abilities of sensing force and holding cell stably. A prototype is fabricated, tested, and applied in crab egg embryo microinjection. Experiments results reveal the promising performance of the fabricated MFC-based microforce sensor in microinjection application.

3.1 Introduction

Biological cell injection is crucial in modern medicine with various applications, e.g., in-vitro fertilization (IVF), gene injection, intracytoplasmic sperm injection (ICSI), and drug development [7]. Up to now, the process of single cell microinjection is mostly performed manually [6, 23]. Normally, it takes several months to train an operator to be skillful to fulfill such operation, that is a time-consuming work [10]. Owing to the rigorous demands to realize a successful injection including longtime attention, patience, and experience, even a skilled worker can only achieve a low success rate around 15% [15].

On the contrary, robotic microinjection of biological cell is an important approach for modern biological applications which demand high precision and throughput. Microforce sensor is an important device to realize a precision injection of biological cells. Depending on the operation principle, the popular microforce sensors can be categorized as capacitive type, piezoelectric type, and piezoresistive type, etc. The comprehensive survey on the most recent development of microforce sensors is conducted in the literature [19].

In particular, piezoelectric microforce sensors have drawn much attention of researchers. As a well-known piezoelectric material, lead zirconate (PZT) has been widely adopted as actuators and sensors [13]. Nevertheless, PZT consists of fragile ceramic material, that is not suitable for most flexible applications (e.g., cell microinjection). Polyvinylidene fluoride (PVDF) is another renowned piezoelectric material. PVDF exhibits several remarkable characteristics in terms of high sensitivity, large

Q. Xu, *Micromachines for Biological Micromanipulation*,
https://doi.org/10.1007/978-3-319-74621-0_3

sensing area, highly thin, wide frequency response range, and low cost [2]. Furthermore, given the same applied force, the PVDF sensor is able to produce a ten times larger voltage output than PZT sensor.

Currently, most of existing PVDF microforce sensors have been constructed as a cantilever structure; i.e., the PVDF film functions as a cantilever and a needle attached on its free end. The PVDF sensor is then mounted on a micromanipulator to perform cell injection task [15]. Such sensor can provide a sensitivity in microforce scale. Yet, in cell microinjection application, the sensor structure possesses several disadvantages. (a) The injection force is difficult to be transferred from PVDF cantilever to the cell, because the PVDF is of highly flexible. (b) The bending deformation of the free end of PVDF film produces a larger puncturing wound to the cell than a rigid injector. (c) The translation of the micromanipulator can cause vibration noise to the microforce sensor.

Recently, other forms of the PVDF microforce sensors in terms of simply supported and fixed-simply supported PVDF film structures have been introduced dedicated to cell microinjection applications [23, 25]. In addition, the idea of designing 3-DOF PVDF microforce sensor as cell holding device has been reported in [4]. However, such sensor structures cannot realize a stable cell immobilization, and a low efficiency of cell injection is resulted. In addition, as the cell moves slightly when a contact force is exerted, the cell injection accuracy is deteriorated. Moreover, as the cell is not immobilized well, the inclined angle between the penetration needle and the cell surface varies during the injection process. As a result, the force sensor with simply supported structure cannot provide an accurate and stable one-dimensional force measurement.

In addition, as a relatively new piezoelectric material, macrofiber composites (MFCs) have been invented in 1996 by NASA. It has been commercialized by Smart Material Corporation since 2002 [18]. Generally, the MFC is made of rectangular piezoceramic rods which are sandwiched between layers of epoxy adhesive, interdigitated electrodes, and Kapton polyimide film. It is of low cost with relatively compact construction. With such a new structure, MFC overcomes the disadvantages of monolithic piezoceramic sensors, such as brittleness and lack of flexibility. Furthermore, when the same force is applied, the distinct interdigitated electrodes of MFC create a higher voltage output than conventional monolithic piezoceramic materials. The reason lies in that such structure can transfer the voltage straightly from the ribbon-shaped rods. MFC has been utilized extensively in various fields such as structural health monitoring, aerospace structures, vibration/noise control of aircraft, and automobiles [3]. However, it is seldom used in microsensor design dedicated to cell microinjection tasks.

In the following parts, a new PVDF-MFC film microsensor based on fixed-guided beam structure is introduced to overcome the disadvantages of present PVDF force sensors [20]. In addition to measure the 1-DOF cell microinjection force, the developed force sensor also functions as a cell holding device in a robotic cell microinjection system.

3.2 Mechanism Design of the Microforce Sensor

To measure the 1-DOF microinjection force and immobilize the cell at the same time, a multistage compound parallelogram flexure (MCPF) [24] is adopted to produce a pure uniaxial translational motion along with large translational range. Generally, a MCPF is composed of N compound parallelogram flexure (CPF).

Figure 3.1a illustrates an MCPF with $N = 1$. It is constructed by four fixed-guided leaf flexures. With an external force F_x is exerted, the MCPF is deflected to create a pure translational motion as illustrated in Fig. 3.1a. Each leaf flexure undergoes a combined force F and moment M as shown in Fig. 3.1b. Concerning the boundary condition, the rotation angle for the end of the flexure is zero to produce a pure translational motion. In view of the boundary conditions as depicted in Fig. 3.1b, the following relations are derived.

$$\frac{Fl^2}{2EI} - \frac{Ml}{EI} = 0 \tag{3.1}$$

$$\frac{Fl^3}{3EI} - \frac{Ml^2}{2EI} = \delta_x \tag{3.2}$$

where E is Young's modulus for the material, l is the flexure length, δ_x denotes the transverse displacement of the flexure, $F = F_x/2$ denotes the force applied on one leaf flexure, and $I = bh^3/12$ is the inertial moment of cross-sectional area. Parameters h and b describe the in-plane width and out-of-plane thickness of the flexure, respectively, as given in Fig. 3.1c.

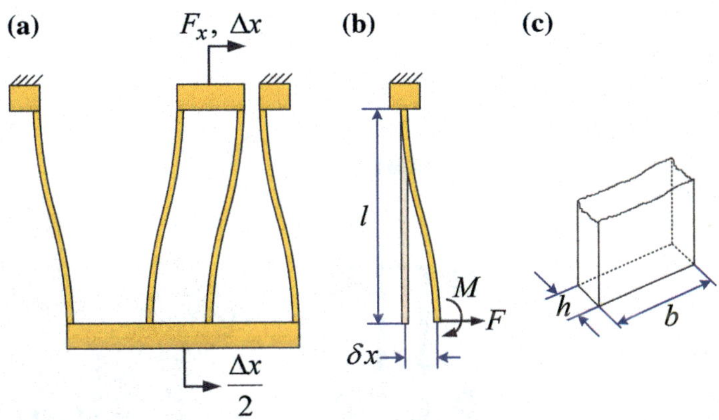

Fig. 3.1 Mechanism and deformation of a compound parallelogram flexure (CPF). **a** Deformation of a typical CPF; **b** deformation of one flexure of CPF; **c** parameters of one flexure

Taking into account Eqs. (3.1) and (3.2), we can obtain

$$F = \frac{2M}{l} \tag{3.3}$$

$$\delta_x = \frac{Fl^3}{12EI} \tag{3.4}$$

Because the four flexures are designed to have the same length l, we can derive that $\delta_x = \Delta_x/2$, where Δ_x represents the uniaxial translation of the CPF.

Figure 3.2 shows the CAD model of the proposed 1-DOF force sensor. Generally, the fixed-guided leaf flexures in the MCPF mechanism can be fabricated out of metal, e.g., Al-7075 [24]. Nevertheless, the metal MCPF's stiffness is too large to be applied as a sensitive microforce sensor. To decrease the stiffness for the MCPF mechanism and to generate a high sensitivity, the leaf flexures are formed by using six PVDF films (model: LDT1-028K, from Measurement Specialties, Inc.) which are denoted by green parts in Fig. 3.2b and two sets of MFC films (model: P2-2807 and P2-2814, from Smart Material Corp.) which are represented by red parts in Fig. 3.2b. These films are much more flexible than any metal materials and have a potential to function as effective microforce sensor.

In order to examine the performance of different MFC materials, two MFC 2807 films and two MFC 2814 films are adopted in sequence in the 1-DOF microforce sensor for experimental testing. Moreover, with the presented design, the performance of the PVDF film and MFC film can be easily compared with each other. To hold the cells firmly, a micropipette holder (model: W4 65-0013, from Warner Instruments Corp.) is used which is depicted by a black cylinder in Fig. 3.2b. In addition, considering that the cell microinjection process will be observed by an inverted microscope, the location of the micropipette holder is designed such that it is as close as possible to the camera lens.

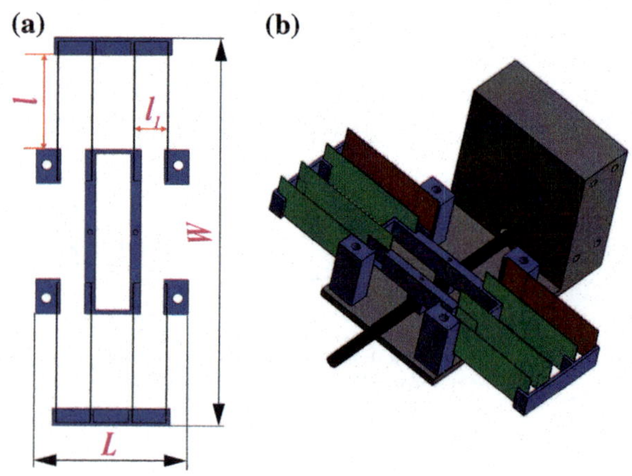

Fig. 3.2 CAD model of the 1-DOF microforce sensor. **a** Top view; **b** 3D view

Table 3.1 Main parameters of the microforce sensor mechanism

Parameter	Value (mm)
L	44
W	114
l	25
l_1	10
b_{PVDF}	16
b_{MFC}	10
h_{PVDF}	0.157
h_{MFC}	0.3

As illustrated in Fig. 3.2, the fixtures are custom-designed to hold these films steadily via adhesive (model: epoxy F-05, from Alteco, Inc.). The main parameters (L, W, l, l_1, b, and h) of the sensor mechanism are denoted in Figs. 3.1b, c, and 3.2a. The parameter values are given in Table 3.1.

3.3 Modeling of the Microforce Sensor

The PVDF and MFC films work as fixed-guided compliant beams in the proposed sensor design. The analytical model of the reported 1-D force sensor is studied in the following.

Firstly, the axial stress is calculated below.

$$\sigma = \frac{F(L-x)\frac{h}{2}}{I} - \frac{M \times \frac{h}{2}}{I} \tag{3.5}$$

where x is the coordinate variable defined in the length direction of the flexure beam.

The aforementioned Eq. (3.5) is held only if the neutral axis of the deflected beam passes through the centroid of the cross-sectional area [16, 21]. Some other expressions of the stress can be referred to the literature [9, 17], which also adopt approximate statements for computational feasibility. The analytical models are developed on the basis of an assumption; i.e., the strain along the width of the beam is zero. However, in the available literature, the influence of the transverse strain has not been quantitatively studied.

The PVDF and MFC films belong to piezoelectric materials. When they are deformed, charges will be produced owing to the direct piezoelectric effect [1, 11]:

$$Q = \int_0^l d_{31}\sigma \mathrm{d}A \tag{3.6}$$

where $A = w \times h$ is the cross-sectional area, d_{31} is the piezoelectric coefficient of PVDF or MFC film, and Q is the charge generated by the PVDF or MFC film.

Inserting Eqs. (3.3) and (3.5) into Eq. (3.6), the charge produced by the fixed-guided beam of piezoelectric film can be calculated as:

$$Q = 0 \tag{3.7}$$

The foregoing result is confirmed by performing finite element analysis (FEA) simulation. For illustration, the simulation result with two MCPFs ($N = 2$) is shown in Fig. 3.3. It reveals that for each flexure beam, the following condition always holds. That is, when one part is under tension, the other part is under compression on one side of the fixed-guided beam. As a result, the strain of the two parts divided by the inflection point, i.e., the minimum strain point in dark blue, is totally antisymmetry along the length direction. Specifically, the MFC/PVDF films produce the voltages (with the same magnitude and inverse polarity) along the length direction, which leads to zero charge along the length direction after the integration operation based on Eq. (3.6). Similar analysis treatment is adopted in the literature [12, 14], in which the pseudo-rigid-body (PRB) model is used and the fixed-guided beam is divided into two identical structures.

But the experimental study conducted later shows that the charge signal is not zero. Instead, the signals are usable for the measurement of the microforce in cell microinjection process. Indeed, the signals are originated from the transverse strain, that is induced by the combined force and moment owing to the Poisson effect. Furthermore, these signals are consistent in terms of the positive and negative values. Thus, the charge is accumulated in the transverse direction. In the previous work [9, 16, 17, 21], such signals are commonly supposed as zero, which may influence the precision of the force sensors. On the contrary, by quantitatively evaluating the effect of the transverse strain, the force measurement is implemented based upon the transverse strain in this work. In the future, a more detailed process for the analytical modeling is deserved.

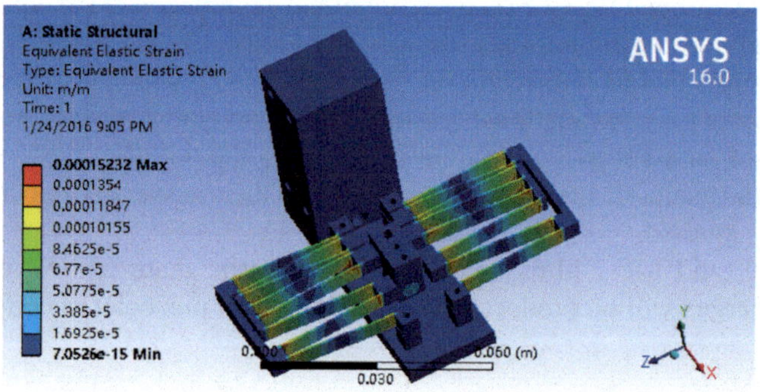

Fig. 3.3 FEA simulation result of the MCPF mechanism with fixed-guided beams

3.4 Fabrication and Calibration of the Microforce Sensor

In this part, the fabrication process of the microforce sensor is presented and experimental study is conducted to calibrate the sensor.

3.4.1 *Experimental Setup*

Figure 3.4 shows a photograph of the microforce sensor prototype, which is developed by two MFC films and six PVDF films. For the microforce sensor calibration, a commercially available force sensor (model: GSO-10, from Transducer Techniques Corp.) is employed, which provides a measurement range of 98.1 mN and resolution of 50 μN.

For the calibration, the 1-DOF microforce sensor is fixed onto an XYZ micromanipulator (model: MP-285, from Sutter Instrument Corp.). The GSO-10 force sensor is mounted on an XYZ stage (from Hangzhou Spectrum Laser Photoelectric Technology Co., Ltd.). Once programmed, the XYZ micromanipulator can translate automatically at a given speed and corresponding resolution (0.04–0.2 μm/step). In practice, the voltages produced by the piezoelectric material are influenced by the speed of exerting the force [5], and the force magnitude detected by the piezoelectric microforce sensor is also affected. Thus, it is essential that the micromanipulator moves automatically for reliable operation of the piezoelectric microforce sensor. Due to this reason, the adopted speed in calibration and application should be the same so that the consistency of the piezoelectric microforce sensor can be guaranteed.

In addition, considering that the speed influences the survival rate of the injected cell [15], a suitable approaching speed is particularly important for cell microinjection application. In the calibration and application, the speed is selected as 330 μm/s for

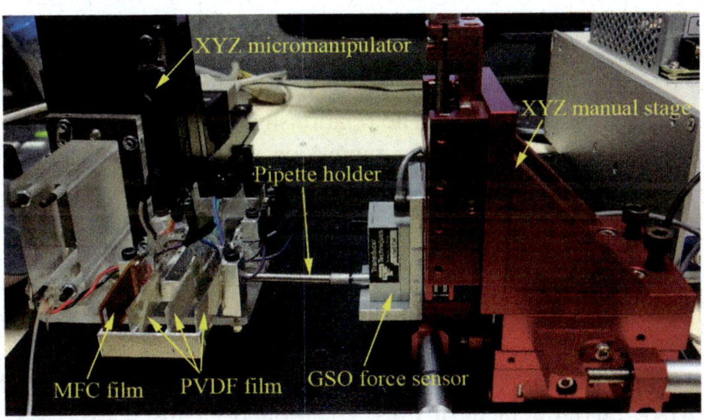

Fig. 3.4 Experimental setup for calibration of the microforce sensor

the designed microforce sensor in this work. The speed is controlled by the built-in software (Multi-link) of the Sutter micromanipulator. Actually, it can work at six speeds of 3700, 745, 330, 235, 130, and 88 μm/s. Preliminary experimental study indicates that the highest speed is too fast to be observed by the used camera in the injection process. Additionally, the largest speed is also prone to damage the injecting pipette and bring some material out of the cell. On the other hand, the lower speeds, i.e., 235, 130, and 88 μm/s, cause a larger deformation before the cell is punctured, that induces bigger damage on the cell. Thus, the other two speeds of 745 and 330 μm/s are tested in the cell microinjection procedure. In this work, the speed of 330 μm/s is chosen as it results in almost 100% success rate in the cell microinjection process.

The XYZ stage is adopted for fixing the GSO force sensor to align the relative positions. Concerning the signal acquisition, a real-time controller (model: cRIO-9075, from National Instruments (NI) Corp.) with NI-9237 and NI-9215 modules is adopted. The bridge input module NI-9237 and NI-9949 RJ-50 screw-terminal accessory are adopted to read the GSO force sensor signals. The simultaneous analog input module NI-9215 and the store-and-hold charge amplifier (model: Smart Charge, from Smart Material Corp.) are used to acquire the output voltages of the PVDF films and MFC films. In addition, NI LabVIEW software is employed to program the signal processing process.

It is noted that majority of previous works using charge amplifiers cannot hold the static value of the voltage produced by piezoelectric material, due to the dynamic nature of the produced signal [8, 22]. Due to this reason, the calibration and application have to be conducted in a short time period, which causes inaccuracy and non-repeatability of the measurement. By contrast, the employed store-and-hold charge amplifier is able to hold the signals without drift for about three minutes, that is very useful in low-speed operation.

During the calibration process, the GSO-10 force sensor is positioned close to the 1-DOF microforce sensor as shown in Fig. 3.4. Then, the microforce sensor is translated close to the GSO-10 force sensor by the XYZ micromanipulator stage with a speed of 330 μm/s. After contact, the microforce sensor is deflected. Meanwhile, the signals generated by the GSO force sensor, MFC films, and PVDF films are acquired by the measurement equipment. To reduce the noise, the signals are low-pass filtered by Labview software. The relationship between the force and voltage signals is studied. Afterward, a linear regression model is obtained for the force-voltage curve with MATLAB software.

3.4.2 Calibration Results

The calibration results for the microforce sensors with PVDF, MFC (P2-2807), and MFC (P2-2814) films are depicted in Figs. 3.5, 3.6, and 3.7, respectively. It is observed from Figs. 3.5a, 3.6a, and 3.7a that the force-voltage curve exhibits a good linearity

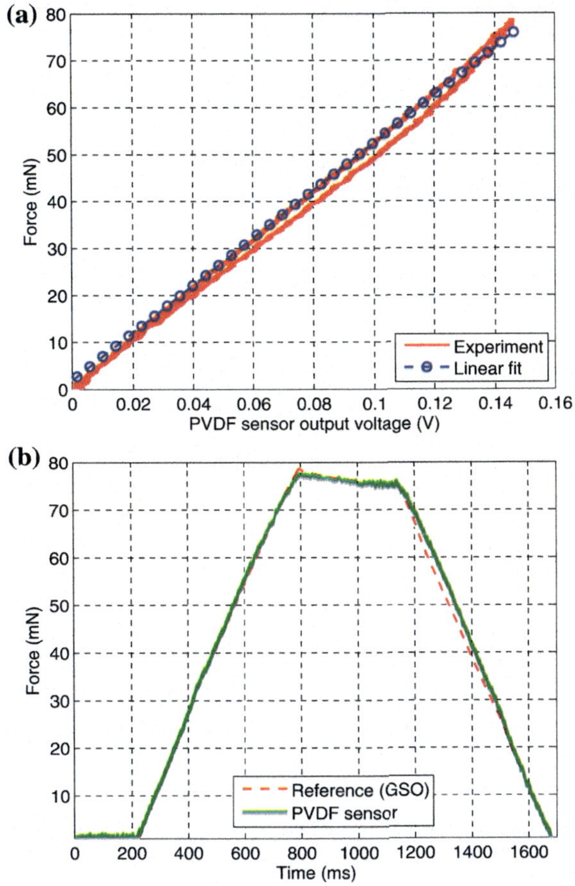

Fig. 3.5 Calibration results of PVDF microforce sensor. **a** Experimental data and fitted linear curve for GSO microforce sensor's output versus PVDF voltage output; **b** comparison between PVDF and GSO microforce sensor outputs

and light hysteresis. In addition, the equations of force–voltage relationships are generated as follows.

$$F_{\mathrm{PVDF}} = 5280 \cdot V \tag{3.8}$$

$$F_{\mathrm{MFC2807}} = 1660 \cdot V \tag{3.9}$$

$$F_{\mathrm{MFC2814}} = 812 \cdot V \tag{3.10}$$

where V denotes the voltage and F_i is the force output of the sensor clement i (i = PVDF, MFC2807, and MFC2814).

During the calibration process, the forces applied on the three kinds of microforce sensors have the same magnitude. Thus, the larger the coefficient of the linear equations (3.8), (3.9), and (3.10), i.e., 5280, 1660, and 812, the less sensitive the sensor. Therefore, the most sensitive microforce sensor is achieved with the MFC-2814 film.

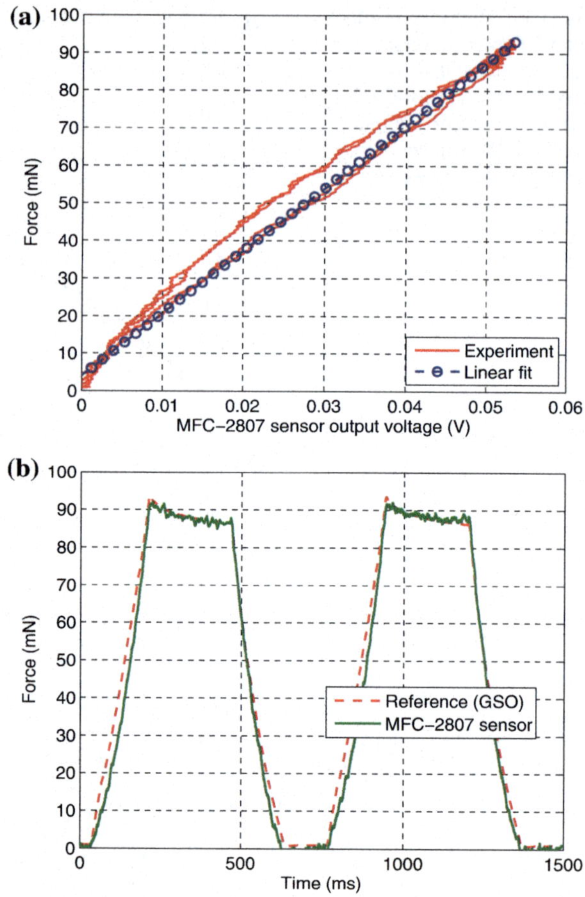

Fig. 3.6 Calibration results of MFC-P2-2807 microforce sensor. **a** Experimental data and fitted linear curve for GSO microforce sensor's output versus MFC-P2-2807 voltage output; **b** comparison between MFC-P2-2807 and GSO microforce sensor outputs

To validate the performance of the three 1-DOF microforce sensors (i.e., PVDF microforce sensor, MFC-2807 microforce sensor, and MFC-2814 microforce sensor), a series of experimental tests are conducted. The experimental results are given in Figs. 3.5b, 3.6b, and 3.7b. We can observe that the calibrated PVDF and MFC microforce sensors exhibit a fine consistency with the GSO commercial force sensor. The discrepancy is attributed to the physical impact and electrical signal drift. The performance testing results of the three microforce sensors imply that MFC-2814 microforce sensor is the most sensitive one, that is consistent with the analytical model prediction.

Based on Eq. (3.10), a sensitivity of 1.23 mV/mN is calculated for the MFC-2814 microforce sensor. The experimental results of resolution testing for the MFC-2814 microforce sensor are depicted in Fig. 3.8. By adopting 2σ (σ is standard deviation) along with 95.44% confidence level, we can see that a resolution around 0.80 mN is produced by the MFC-2814 microforce sensor. In addition, MFC-2814

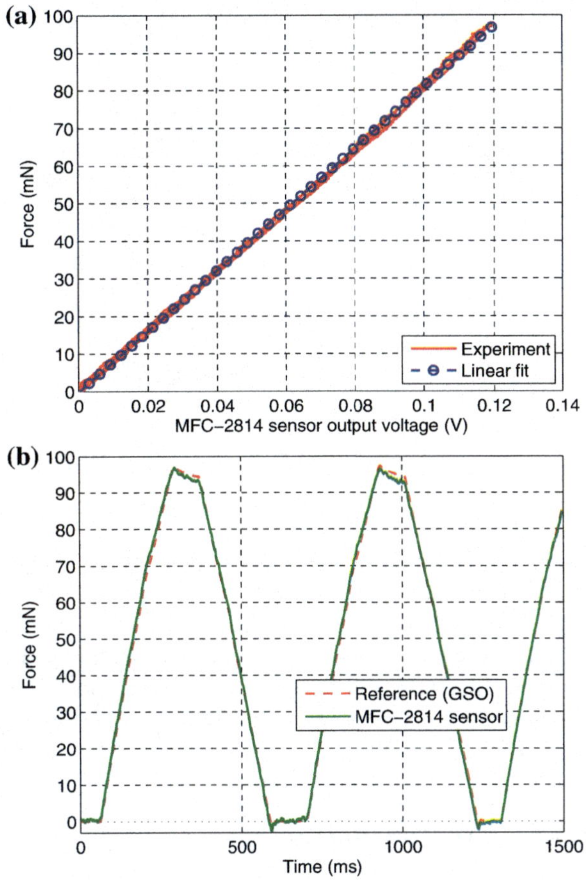

Fig. 3.7 Calibration results of MFC-P2-2814 microforce sensor. **a** Experimental data and fitted linear curve for GSO microforce sensor's output versus MFC-P2-2814 voltage output; **b** comparison between MFC-P2-2814 and GSO microforce sensor outputs

microforce sensor provides a measurement range of 100 mN. Therefore, the MFC-2814 microforce sensor has better performance and it is selected to provide force sensing in the following cell microinjection application.

It is noted that multiple MFC and PVDF films can be cascaded in series to improve the sensitivity for the microforce sensor. Yet, the initial values of the film outputs are commonly not the same in practice. Therefore, a suitable signal processing is required to implement the conceptual design of a cascade microforce sensor.

3.5 Application in Cell Microinjection

In order to verify the performance of the designed 1-DOF microforce sensor in cell microinjection task, crab eggs are adopted as samples in the experimental studies. The size of crab egg is about 1 mm diameter, which is similar to that of Xenopus (frog

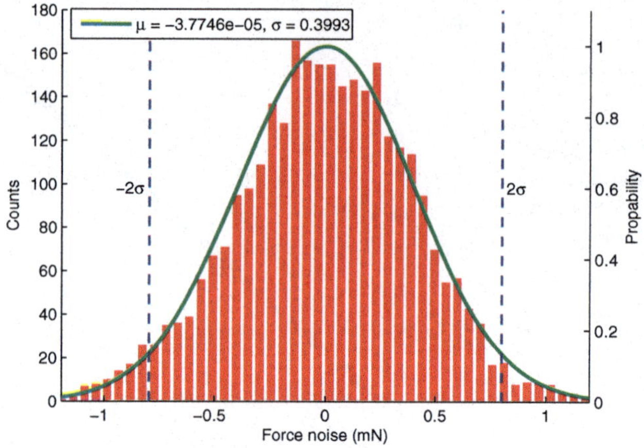

Fig. 3.8 Noise histogram of the MFC-P2-2814 microforce sensor

egg). Xenopus embryos are popularly used in study of embryonic development owing to a number of advantages involving easily identifiable blastomeres, large size of the eggs, and the ability of withstanding extensive surgical intervention and culture. Xenopus embryos have been commonly adopted in cell microinjection research. Xenopus injection will be executed in the future application. The current testing is performed to verify the proposed conceptual design of the microforce sensor.

3.5.1 Experimental Setup

Figure 3.9 depicts the experimental setup in this work. In order to deliver extra materials into a cell and hold the cell firmly, a pico-injector (model: PL1-100A, from Warner Instruments Corp.) along with high-pressure nitrogen is employed. The pico-injector provides the functions of both injecting and holding. To implement automated cell injection, the injection motion is offered by an XYZ micromanipulator (model: MP-285, from Sutter Instrument Corp.). Given the trajectory and speed, the micromanipulator is controlled by an associated controller (model: MPC-200, from Sutter Instruments Corp.) through Multi-link software. The operation is observed with an inverted microscope (model: IX81, from Olympus Corp.) for monitoring the injection process. As depicted in Fig. 3.9b, the injecting micropipette is pulled by using a micropipette puller (model: P-1000, from Sutter Instrument Corp.). The operation is conducted in accordance with the instruction of pipette pulling for injecting a Xenopus. The holding micropipette is forged to firmly hold the cell and to protect the cell from sharp blade of the pipette edge. Furthermore, the front parts of both injecting micropipette and holding micropipette are forged and bent. Then, they are aligned in horizontal direction to realize the pure uniaxial force measurement.

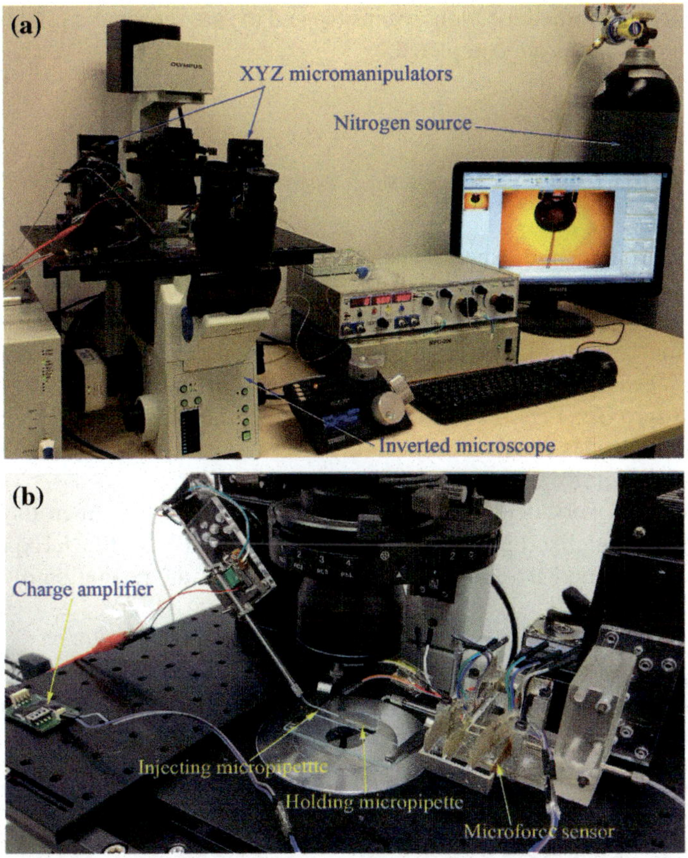

Fig. 3.9 **a** Overall view and **b** close-up view of experiment setup for crab egg microinjection

The experimental study is performed based on the following procedures.

1. Dyed water is absorbed into the injecting pipette and holding pipette. Indeed, the dyed water is employed as buffer liquid in the holding pipette.
2. The microscope focus and positions of the injecting and holding pipettes are adjusted by XYZ micromanipulators so that they are displayed clearly in the microscope's field of view.
3. The two pipettes are aligned close to each other, and the locations of the two micromanipulators are recorded by using the Multi-link software. In this way, the experiments can be repeated multiple times.
4. The holding pressure is increased by controlling the pico-injector in order to hold the crab egg.
5. The crab egg is pierced at a speed of 330 μm/s. The piercing process is stop once the cell is penetrated, which is detected by monitoring the force signal with the designed force sensor. Afterward, the injecting liquid is delivered.

6. Return to the recorded locations. Steps (4) and (5) are repeated until the required injection operations are completed.

3.5.2 Results and Discussions

Figure 3.10 illustrates the experimental results of the crab egg microinjection. As shown in Fig. 3.10a, When the injecting pipette contacts the crab egg initially, the force signal rises rapidly which is displayed in Fig. 3.11. Then, translating the injecting pipette forward, the deformation and the force signal rise as shown in Fig. 3.10b. When the crab egg is punctured, there is a dramatic drop of the force signal down to zero (see Figs. 3.10c and 3.11), as little resistance is presented. Experimental result indicates that the designed force sensor can detect the injection process perfectly and faithfully measures the puncturing force.

The foregoing work demonstrates the feasibility of the new microforce sensor that is designed with fixed-guided beams based on MFC/PVDF films. Experimental results reveal that the MFC microforce sensor provides the resolution of 0.80 mN,

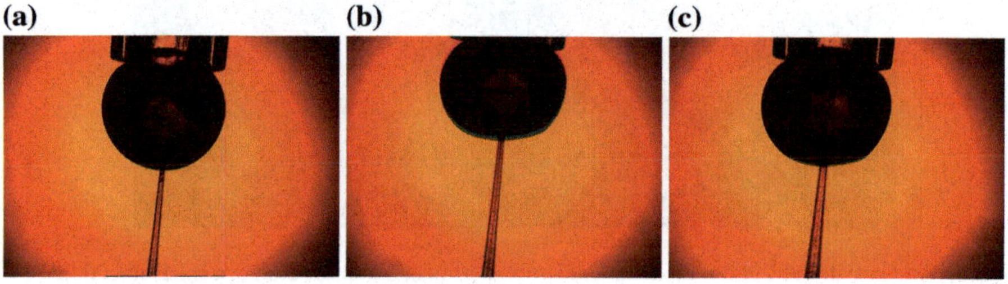

Fig. 3.10 Snapshots of microinjection for a crab egg. **a** Initial contact is made between the injecting pipette and crab egg; **b** crab egg is obviously deformed before punctured; **c** crab egg is punctured

Fig. 3.11 Output signal of the designed microforce sensor during the cell microinjection process, which shows a puncturing force about 27 mN

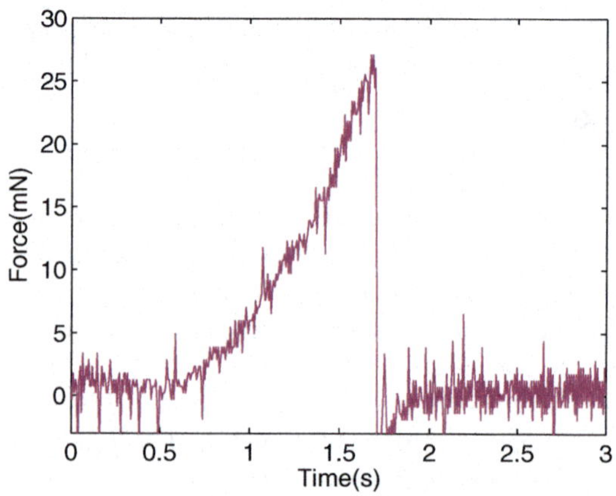

sensitivity of 1.23 mV/mN, and measurement range of 100 mN. The reported sensor not only can steadily detect the pure 1-DOF cell microinjection force, but also can stably hold the cell. Moreover, the introduced concept design of a 1-D MFC/PVDF microforce sensor can be easily expanded for multi-axis force sensor design targeting at pertinent applications.

3.6 Conclusion

In this chapter, a 1-DOF microforce sensor is designed, fabricated, calibrated, and applied to crab egg cell microinjection process. The microforce sensor can detect the pure injection force based on the MCPF mechanism. The microforce sensor not only can measure the cell injection force, but can immobilize the cell firmly. The performance of the microforce sensors using PVDF film and two types of MFC films are obtained for comparison. Results show that the MFC film exhibits promising performance in cell microinjection. Experimental studies of cell microinjection demonstrate that the designed microforce sensor faithfully monitors the procedure of cell microinjection and stably measures the puncturing force of 27 mN for a crab egg. Furthermore, the transverse strain effect is firstly measured quantitatively thanks to the designed sensor structure on the basis of fixed-guided beams, that supplies a reference for the further study.

References

1. An, Z., Xu, M., Luo, Y., Wu, C.: Active vibration control for a large annular flexible structure via a macro-fiber composite strain sensor and voice coil actuator. Int. J. Appl. Mech. **7**(04), 1550066 (2015)
2. Chuang, K.C., Huang, K.C., Liou, H.C., Ma, C.C.: Investigation of transient behavior of a cantilever plate based on elastic impact loading history detection using polyvinylidene fluoride film sensors. IEEE Sens. J. **16**(5), 1565–1574 (2016)
3. Dan, C., Malinowski, P., Kudela, P., Opoka, S., Radzienski, M., Mieloszyk, M., Ostachowicz, W.: Validation of macro fiber composites for strain measurements in structural health monitoring applications of complex aerospace structures. In: Proceedings of The 9th International Conference on Structural Dynamics (EURODYN 2014), pp. 2127–2134 (2014)
4. Faramarzi, S., Ghanbari, A., Chen, X., Wang, W.: A pvdf based 3d force sensor for micro and nano manipulation. In: Proceedings of 2009 IEEE International Conference on Control and Automation, pp. 867–871 (2009)
5. Fung, C.K., Elhajj, I., Li, W.J., Xi, N.: A 2-d pvdf force sensing system for micro-manipulation and micro-assembly. In: Proceedings of IEEE International Conference on Robotics and Automation (ICRA'02), vol. 2, pp. 1489–1494 (2002)
6. Ghanbari, A., Horan, B., Nahavandi, S., Chen, X., Wang, W.: Haptic microrobotic cell injection system. IEEE Syst. J. **8**(2), 371–383 (2014)
7. Huang, H., Sun, D., Mills, J.K., Cheng, S.H.: Automatic suspended cell injection under vision and force control biomanipulation. In: Proceedings of IEEE International Conference on Robotics and Biomimetics (ROBIO 2007), pp. 71–76 (2007)

8. Kim, J., Janabi-Sharifi, F., Kim, J.: Haptic feedback based on physically based modeling for cellular manipulation systems. In: Proceedings of International Conference on Human Haptic Sensing and Touch Enabled Computer Applications, pp. 661–667 (2008)
9. Liu, Y., Yu, P., Wang, Y., Dong, Z., Xi, N.: The design and development of a mirco-force sensing device. In: Proceedings of IEEE International Conference on Robotics and Biomimetics (ROBIO 2007), pp. 77–81 (2007)
10. Lu, Z., Chen, P.C., Nam, J., Ge, R., Lin, W.: A micromanipulation system with dynamic force-feedback for automatic batch microinjection. J. Micromech. Microeng. **17**(2), 314 (2007)
11. Meitzler, A., Tiersten, H., Warner, A., Berlincourt, D., Couqin, G., Welsh III, F.: IEEE Standard on Piezoelectricity (1988)
12. Midha, A., Bapat, S.G., Mavanthoor, A., Chinta, V.: Analysis of a fixed-guided compliant beam with an inflection point using the pseudo-rigid-body model concept. J. Mech. Robot. **7**(3), 031007 (2015)
13. Nguyen, T.D., Deshmukh, N., Nagarah, J.M., Kramer, T., Purohit, P.K., Berry, M.J., McAlpine, M.C.: Piezoelectric nanoribbons for monitoring cellular deformations. Nat. Nanotechnol. **7**(9), 587–593 (2012)
14. Pei, X., Yu, J., Zong, G., Bi, S.: An effective pseudo-rigid-body method for beam-based compliant mechanisms. Precis. Eng. **34**(3), 634–639 (2010)
15. Pillarisetti, A., Anjum, W., Desai, J.P., Friedman, G., Brooks, A.D.: Force feedback interface for cell injection. In: Proceedings of the First Joint Eurohaptics Conference and Symposium on Haptic Interfaces for Virtual Environment and Teleoperator Systems, pp. 391–400 (2005)
16. Pillarisetti, A., Pekarev, M., Brooks, A.D., Desai, J.P.: Evaluating the effect of force feedback in cell injection. IEEE Trans. Autom. Sci. Eng. **4**(3), 322 (2007)
17. Shen, Y., Winder, E., Xi, N., Pomeroy, C.A., Wejinya, U.C.: Closed-loop optimal control-enabled piezoelectric microforce sensors. IEEE/ASME Trans. Mechatron. **11**(4), 420–427 (2006)
18. Sodano, H.A.: Macro-fiber composites for sensing, actuation and power generation. Ph.D. thesis, Virginia Polytechnic Institute and State University (2003)
19. Wei, Y., Xu, Q.: An overview of micro-force sensing techniques. Sens. Actuators, A **234**, 359–374 (2015)
20. Wei, Y., Xu, Q.: Design of a PVDF-MFC force sensor for robot-assisted single cell microinjection. IEEE Sens. J. **17**(13), 3975–3982 (2017)
21. Xi, W., Zhong, H.: Bio-manipulation probe integration with micro-force sensor. In: Proceedings of 3rd IEEE International Conference on Nano/Micro Engineered and Molecular Systems (NEMS 2008), pp. 393–396 (2008)
22. Xie, Y., Sun, D., Liu, C.: Penetration force measurement and control in robotic cell microinjection. In: Proceedings of 2009 IEEE/RSJ International Conference on Intelligent Robots and Systems, pp. 4701–4706 (2009)
23. Xie, Y., Sun, D., Tse, H.Y.G., Liu, C., Cheng, S.H.: Force sensing and manipulation strategy in robot-assisted microinjection on zebrafish embryos. IEEE/ASME Trans. Mechatron. **16**(6), 1002–1010 (2011)
24. Xu, Q.: New flexure parallel-kinematic micropositioning system with large workspace. IEEE Trans. Rob. **28**(2), 478–491 (2012)
25. Zhou, Y., Liu, S., Yang, C., Liu, C., Xie, Y.: Design of a microforce sensor based on fixed-simply supported beam: towards realtime cell microinjection. In: Proceedings of 2015 IEEE International Conference on Cyber Technology in Automation, Control, and Intelligent Systems (CYBER), pp. 1080–1084 (2015)

Chapter 4
Design and Control of a Piezoelectric-Driven Microinjector

Abstract This chapter presents the design, fabrication, and testing of a piezo-driven cell microinjection system with force sensing and control. By using piezoresistive sensors for measuring the cell penetration force and micropipette position in real time, the developed cell microinjection system enables high operation speed, high success rate, and high survival rate. The effectiveness of the microinjection system is verified by penetrating zebrafish embryos with both position control and force control through experimental study. Results indicate that the force control produces a higher survival rate than position control. A smooth position and force switching control is proposed to mitigate the oscillation during the transition between the two control objectives. The experimental results quantitatively demonstrate the superiority of force control over conventional position control for biological cell microinjection.

4.1 Introduction

Biological cell microinjection is the process of delivering exogenous materials into cells, which is a common operation in biological experiments. Such operation has been widely applied in genetic engineering, intracytoplasmic sperm injection (ICSI), DNA therapy, or other biomedical areas. Comparison study of the human reference genome shows that approximately 70% of human genes have at least one obvious zebrafish orthologue [7]. Owing to this reason, zebrafish embryo has been extensively used in disease treatment and drug discovery research.

To overcome the weakness of manual operation, automated injection technologies have been widely investigated over the last few decades. For example, a microrobotic system with a visual servo control has been developed for automatic cell injection in [18], which greatly reduces the manipulation time and improves the injection success rate. Optical tweezers have been introduced to conduct cell patterning for the execution of the required operation [30]. In addition, a comprehensive survey of recent development of automated microinjection systems has been presented in [14].

There are two main criteria in cell injection, i.e., success rate and survival rate. In the existing automatic cell injection systems, image-based visual servoing is the dominant control approach [1, 11, 33]. However, due to the time delay introduced by

Q. Xu, *Micromachines for Biological Micromanipulation*,
https://doi.org/10.1007/978-3-319-74621-0_4

the image acquisition and processing process, the pipette location that is extracted by an image processing algorithm is not fully synchronous with the cell status. Furthermore, the vision-based feedback alone is difficult to determine whether the pipette has slipped off or pierced into the cell during the operation. As a result, a low success rate of cell injection is produced. On the contrary, it has been shown that a higher success rate will be achieved with force feedback intervened [15]. In the literature, the cell force model has been developed by bulk-scale methods, which rely on the average value of hundreds or thousands of cells. However, the individual cell's characteristics cannot be reflected with the cell group [17]. For an accurate measurement of the exerted force, a microinjection prototype with the integration of a piezoresistive microforce sensor was developed for zebrafish embryos [13]. It can synchronize the position control with force profile during the injection process.

To improve the cell survival rate, it is necessary to minimize the adverse effect of deformation for the zebrafish embryos induced by the injection force. To cope with this problem, piezoelectric actuators (PZTs) have been suggested thanks to with the advantages including fast response speed and high positioning resolution [27]. Piezo-driven micropipette was first introduced into cell injection by Kimura [16], in which the pipette was driven by a serial of piezopules to execute mouse oocytes injection. A survival rate of 80% was achieved in that experiment, while only 16% of the oocytes survived with conventional method. Afterward, several remarkable research works have been conducted with piezo-driven cell injection system [2, 3, 5, 6, 8, 10]. However, all of these works are conducted based on the high-frequency vibration of PZTs. It means that the PZT acts like a vibration drill and it is combined with the other actuators to perform the cell injection tasks. Nevertheless, it has been argued that the high-frequency vibration may change the original status and cause adverse effects for the cell [32].

In this chapter, the PZT is adopted to drive a microinjector with a flexure guiding mechanism to drive the pipette directly for executing cell injection. Up to now, it is unknown to what extent that the force control performs better than the position control in biological cell injection manipulation. In this work, a piezoelectric actuator with a displacement amplifier is designed to drive the injection pipette. Both position and force sensors are integrated into the device to improve the compactness. Comparative studies with position control and force control have been carried out by experimental injection of 100 zebrafish embryos, and some conclusions are derived. Moreover, a scheme of smooth position/force switching control is proposed to reduce the damage caused on the cell.

4.2 Mechanism Design of the Piezo-Driven Cell Microinjector

In this section, the piezo-driven cell microinjector involving the injector and position and force sensors is designed.

Usually, the commercial piezoelectric actuators (PZTs) can only provide a constrained output displacement (e.g., 32 μm in this work). Such a limited stroke is insufficient to pierce zebrafish embryos which have the diameter around 600–1200 μm. Thus, a displacement amplifier is needed to cater for the motion range requirement. Moreover, the injector micropipette should experience a pure in-plane translation and avoid the lateral movement, because the latter motion can induce extra damage to the penetrated cells [19].

In the literature, various displacement amplifiers have been presented for piezoelectric actuators [28]. In this work, a piezo-driven flexure-based injector is designed with a bridge-type displacement amplifier as shown in Fig. 4.1. The direction of output motion for the PZT is orthogonal to the injection direction. In order to achieve a high-precision in-plane positioning motion, two leaf flexures in the outside frame

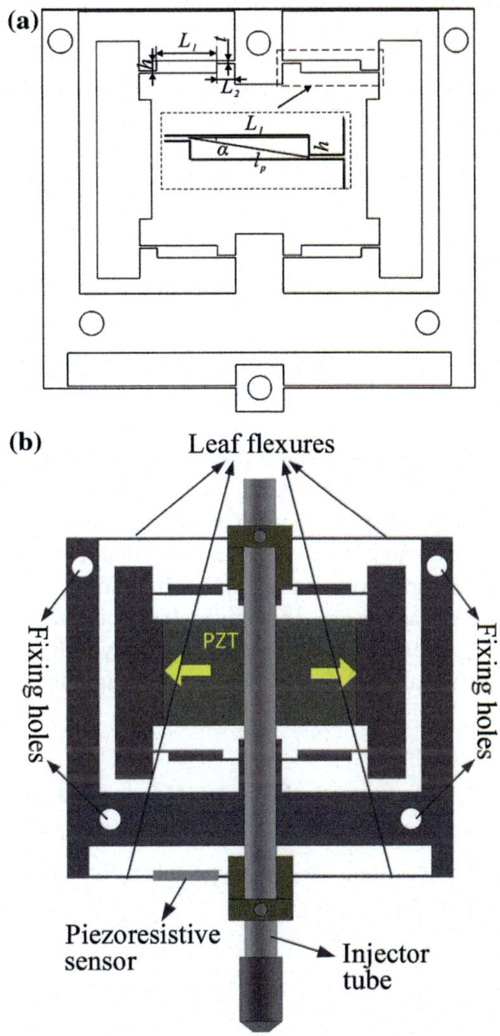

Fig. 4.1 **a** Schematic diagram and **b** CAD model of the flexure-based injector, which is driven by a PZT with a bridge-type displacement amplifier

Table 4.1 Main parameters of the compliant injector

Parameter	Symbol	Value (mm)
Length of amplifier arm	L_1	9.0
Length of amplifier flexure	L_2	4.5
Width of amplifier arm	h	1.0
Width of amplifier flexure	t	0.5

are used to guide the linear output motion of the injector. As compared with a parallelogram-based amplifier system, the adopted displacement amplifier reduces the number of guiding flexures and simplifies the structure design, which enables a compact physical design with cost reduction on fabrication.

To facilitate the analytical modeling, it is assumed that each flexure hinge experiences a one-degree-of-freedom (1-DOF) rotational compliance which arises from the bending deformation, and other elements are all rigid bodies. Then, the displacement output in the injection direction can be calculated by [29]:

$$\Delta y = l_p \sin\alpha - \sqrt{l_p^2 \sin^2\alpha - \Delta x^2 - 2l_p \cos\alpha \Delta x} \tag{4.1}$$

In addition, the amplification ratio of the bridge-based amplifier can be calculated as follows.

$$R = \frac{\Delta y}{\Delta x} = \frac{l_p \sin\alpha - \sqrt{l_p^2 \sin^2\alpha - \Delta x^2 - 2l_p \cos\alpha \Delta x}}{\Delta x} \tag{4.2}$$

For illustration, the injector mechanism is designed with the main parameters given in Table 4.1. For a driving displacement of 32 μm, analytical model predicts that an output displacement of 350 μm can be produced. Moreover, simulation study with finite element analysis (FEA) is conducted with ANSYS software. The simulation results are shown in Fig. 4.2, which indicate an output displacement of 398 μm.

We can observe that the analytical model result is 13.7% smaller than the simulation result. The discrepancy is mainly induced by the assumption introduced in the analytical model, which only considers the compliance of the flexure hinges [12].

4.3 Prototype Fabrication and Calibration

In this section, the prototype fabrication is presented and the sensor calibration is conducted by experiments.

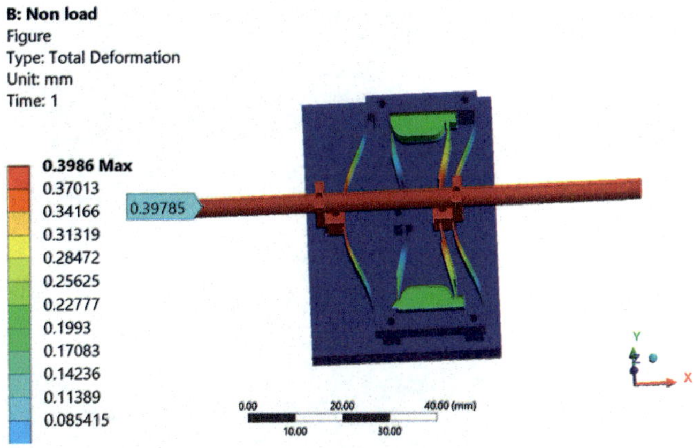

Fig. 4.2 FEA simulation results of the injector mechanism

4.3.1 Prototype Fabrication and Experimental Setup

The CAD model of the flexure-based cell injector along with PZT actuator and strain-gage sensors is shown in Fig. 4.3. A prototype of the flexure-based microinjector mechanism is developed as depicted in Fig. 4.4. It is fabricated with Al-7075 alloy using wire electrical discharge machining (EDM) process. The injector is driven by a piezoelectric actuator (PEA) (stroke: 32 μm, model: P-885.91, from Physik Instrumente, Co.). The PEA is actuated by a high-voltage amplifier (model: EPA-104, from Piezo System, Inc.), by which the voltage will be enlarged by ten times to drive the flexure-based microinjector and deliver a maximum displacement of 200 μm. Moreover, FEA simulation study shows that an output displacement of 301 μm is achieved when the mass of the injector is applied as a payload in the vertical out-of-plane direction. As compared with FEA simulation result, the experimental result

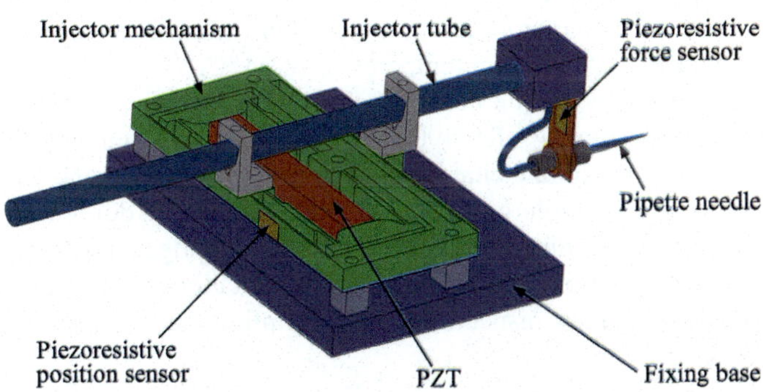

Fig. 4.3 CAD model of the developed piezo-driven injector with strain-gage position and force sensors

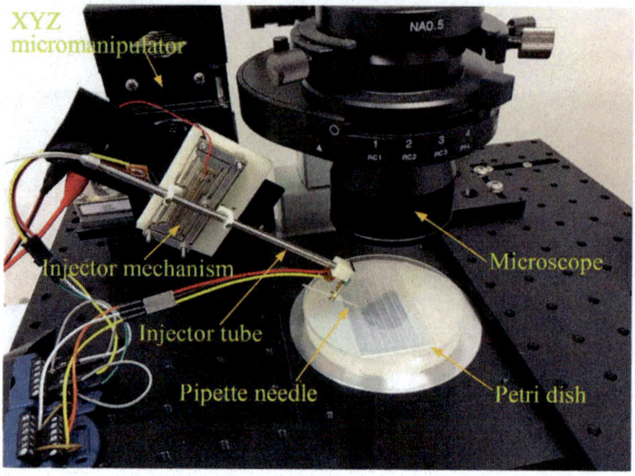

Fig. 4.4 Experimental setup of the cell injection system

is 33% smaller. The discrepancy is mainly attributed to the fabrication errors of the mechanism parameters.

In order to measure the output displacement of the microinjector and the force applied on the injection pipette, two half-bridge circuits are constructed by using piezoresistive strain gages (model: TP-3.8-350, from Bengbu Tianguang Sensor, Ltd.). The commercial laser displacement sensor (model: LK-H055, from Keyence Corp.) and load cell (model: GSO-10, from Transducer Technology, Inc.) are used for the calibration of the custom-built position and force sensors, respectively. In addition, a thick-walled glass capillary (borosilicate glass B100-50-10, from Sutter Instrument Co.) is pulled using micropipette puller (model: P-1000, from Sutter Instrument Co.) to generate an appropriate outer diameter (3 μm) for the injection pipette. An inverted microscope (model: IX81, from Olympus, Inc.) with CCD camera is used for monitoring the injection process. The microscope is equipped with an XY stage for positioning the petri dish. Zebrafish embryos are placed in parallel V-grooves, which are made from the 1.5% agarose gel in a petri dish. The microinjector can move in plane for injecting the embryos in each groove, which simplifies 3D movement into 2D planar motion. The calibration and experiments with zebrafish embryos are conducted at the room temperature of 24 °C.

The control algorithm for the cell microinjection system is implemented with a real-time controller (model: cRIO-9022, from Natural Instruments (NI), Corp.) integrated with a reconfigurable chassis (model: cRIO-9118, from NI Corp.) which contains a field-programmable gate array (FPGA) module. Additionally, NI-9263 analog output module (16-bit resolution) and NI-9237 bridge analog input module (24-bit resolution) are employed to produce the excitation voltage and obtain the piezoresistive sensor signals, respectively. The sampling rate is selected as 4 kHz in the experiments.

4.3.2 *Calibration of Position Sensor*

In order to form a half-bridge circuit for the position measurement, two piezoresistive sensors are mounted on the flexure of displacement amplifier, as shown in Fig. 4.4. For the position sensor calibration, the displacement of the pipette is measured by a laser displacement sensor (model: LK-H055, from Keyence Corp.), and the corresponding output value of piezoresistive sensor is recorded. A linear relation between the laser output and strain-gage values is observed, and the sensitivity of the piezoresistive sensor is derived as $-14.42\,\mu$m/mV.

Figure 4.5a indicates that the calibrated piezoresistive sensor matches well with the laser sensor. When a zero voltage input is applied, the noise of the piezoresistive sensor is sampled as shown in Fig. 4.5b. The noise follows a normal distribution as shown in Fig. 4.5c. By adopting 2σ (σ is the standard deviation) of sensor noise as the resolution [26], the resolution of the piezoresistive position sensor can be obtained as $0.054\,\mu$m.

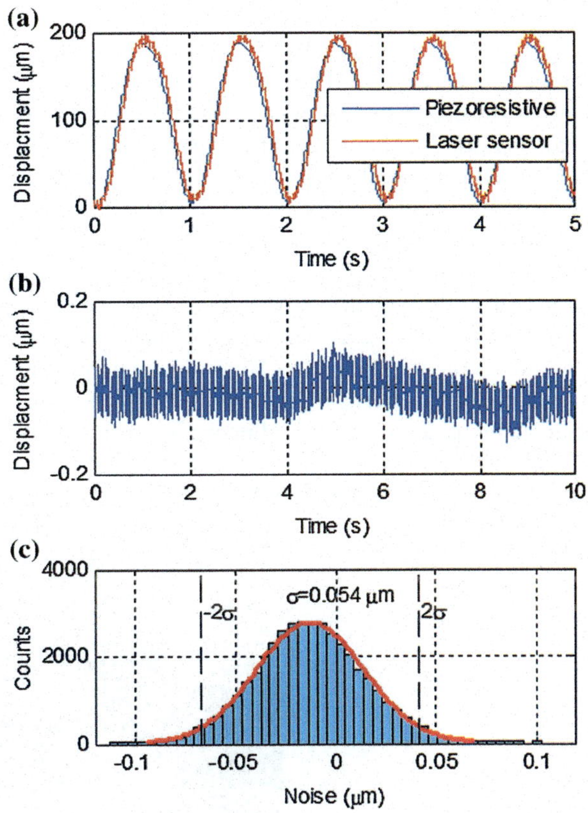

Fig. 4.5 Calibration result of piezoresistive position sensor. **a** Output signals of laser sensor and piezoresistive position sensor; **b** piezoresistive position sensor noise; **c** histogram of piezoresistive position sensor noise

4.3.3 Calibration of Force Sensor

In the cell injection, majority of force sensors are designed with polyvinylidene fluoride (PVDF) film [9, 15, 23, 25], which provides a remarkable dynamic force detection ability. However, the dynamic sensor is not suitable for the sensing of quasi-static signals. Alternatively, with the advantages of higher sensitivity, piezoresistive sensors are used to measure quasi-static force signals in this work. It works based on the piezoresistive effects of silicon semiconductor and measures the change in resistance induced by the accompanied strain.

As shown in Fig. 4.4, two piezoresistive sensors are mounted on a cantilever-beam mylar to construct a half-bridge circuit. It is adopted to measure the injection force in this work. The experimental setup for the calibration of the developed force sensor is shown in Fig. 4.6. In particular, the commercial load-cell force sensor (model: GSO-10, from Transducer Techniques Corp.) with a maximum measurement range of 98.1 mN and resolution of 50 μN is adopted. It is fixed onto a 3-DOF high-precision positioning stage (model: HTCL25-X, from Huntington Optics, Inc.). An approximate linear relationship between the load cell and strain-gage signal values is observed. The sensitivity of piezoresistive sensor is derived as −14.204 mN/mV.

Figure 4.7a shows that the output of the calibrated piezoresistive microforce sensor is consistent with the readings of load-cell sensor. In comparison to PVDF sensors, the disadvantages of piezoresistive sensor lie in the relatively large sensitivity to temperature variation and drift effect. The temperature drift is notable as shown in Fig. 4.7b. By adopting 2σ of sensor noise as the resolution [26], the developed piezoresistive sensor offers a resolution of 0.218 mN, as indicated in Fig. 4.7c.

By inspecting the results in Figs. 4.7c and 4.5c, it is observed that the temperature drift is smaller in the position sensor in comparison with the force sensor using the same half-bridge signal conditioning circuit. The reason may lie in that the temperature on position sensor is more stable than that on force sensor. The displacement amplifier, which is made from aluminum material, has a better thermal conductivity than the flexure material of mylar for the force sensor. As a result, the imperceptible temperature change has less influence on the position strain-gage sensor.

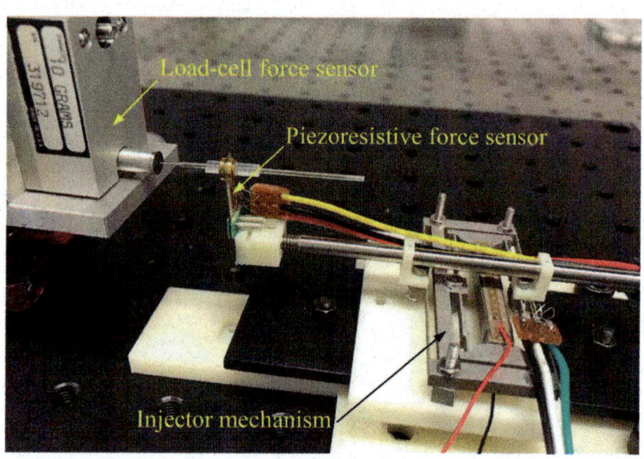

Fig. 4.6 Experimental setup for the force sensor calibration

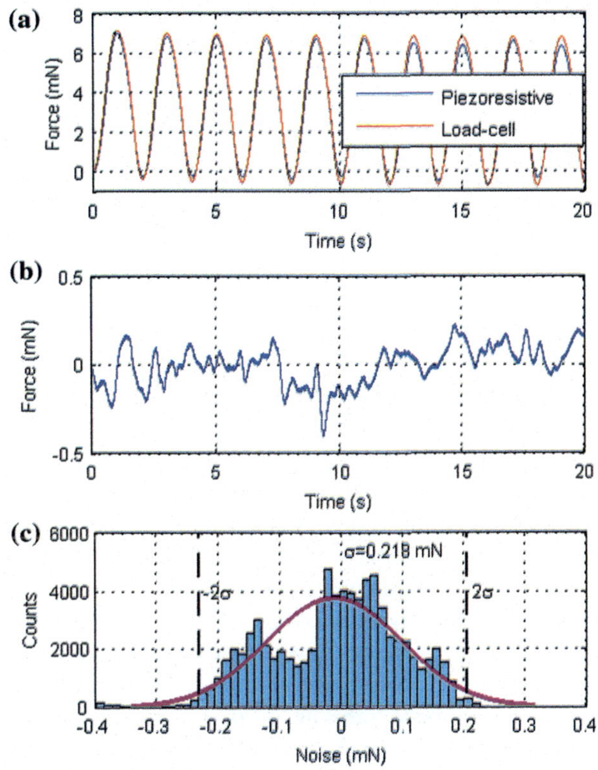

Fig. 4.7 Calibration result of piezoresistive force sensor. **a** Output of load-cell force sensor and piezoresistive force sensor; **b** piezoresistive force sensor noise; **c** histogram of piezoresistive force sensor noise

4.4 Preliminary Experimental Study

In this section, the fundamental position and force controllers are designed and motion planning is presented for cell injection application of the developed cell microinjection system.

4.4.1 Position and Force Controller Design

To compare the performances of cell injection system with the traditional position control and position/force control during cell puncturing process, the same controller structure is adopted. Particularly, in consideration of its popularity and model-free nature, the digital proportional–integral–derivative (PID) control algorithm is employed to construct the position and force controllers for cell injection system. The position/force tracking error is defined as follows:

$$e(t) = x(t) - x_r(t) \tag{4.3}$$

where x and x_r represent the actual and desired position/force values, respectively. Then, the control action is derived below.

$$\begin{aligned} u(t) =& u(t-T) + K_p[e(t) - e(t-T)] + K_i e(t) \\ &+ K_d[e(t) - 2e(t-T) + e(t-2T)] \end{aligned} \tag{4.4}$$

where t represents the time variable, T denotes the sampling time interval, $u(t)$ and $u(t-T)$ are the control actions in the current and previous time steps, respectively. In addition, K_p, K_i, and K_d are the positive control gains.

4.4.2 Motion Planning for Cell Microinjection

For cell microinjection testing, the fabricated injector is mounted on an XYZ micromanipulator (model: MP-285, from Sutter Instrument Corp.) to develop an automated cell injection system. In the experimental study, a batch of zebrafish embryos are immobilized on the V-groove agarose gel in a petri dish, which is placed on the platform of an inverted microscope. The injector tip can move in plane for injecting embryos in each V-groove. After the initialization and calibration of the CCD camera, the relative position between the pipette and target embryo can be determined by means of computer vision algorithm. As controlled by the XYZ micromanipulator, the micropipette translates from the initial position to the target cell firstly with a coarse motion. Once the pipette needle contacts with the embryo, which can be detected by the integrated force sensor, the coarse movement of the micromanipulator is switched to the fine motion of the piezo-driven injector with PID motion control algorithm. In this work, two control schemes are realized for comparison, i.e., the traditional injection sequence with position control only and the improved operation flow with both position and force control.

A predefined position/force trajectory is used as a reference based on the trigger value for pipette puncturing until a dramatic force drop is detected. Then, a time interval is reserved for executing practical injection task, i.e., injecting the desired material. Afterward, a gentle retracting movement is applied under the precision position control to avoid throwing out the injected material by the retracting pipette. Finally, when the piezoelectric actuator returns to its home position, the micromanipulator translates back with a certain distance to prevent the pipette needle from crashing into the embryo during the movement to the next embryo. When the next embryo is translated into the field of view of the camera, the above injection process is repeated. The motion sequences are illustrated in Fig. 4.8.

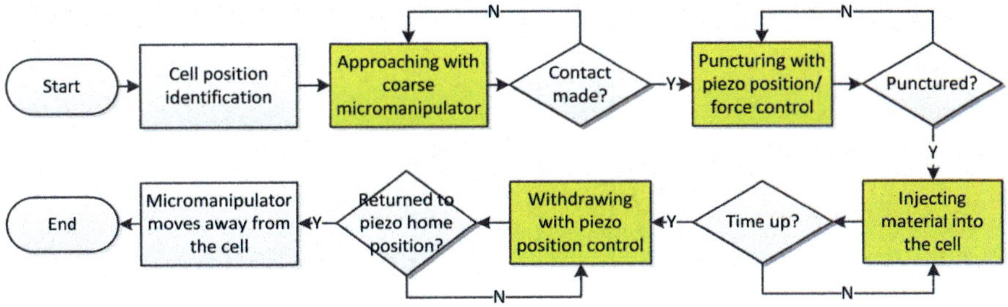

Fig. 4.8 Motion sequences of the automated cell microinjection system

4.4.3 Experimental Study of Cell Microinjection

In practical microinjection operation, majority of unsuccessful cases are induced by the fact that the computer believes that the target cell has been punctured while the pipette needle did not really do it. Actually, with the stand-alone visual feedback, it is difficult to distinguish whether the micropipette has contacted the target embryo or just slipped over it.

For example, two injection scenarios are depicted in Fig. 4.9. Figure 4.9a reveals that the pipette is puncturing the embryo while Fig. 4.9b illustrates the case in which the pipette slips over the embryo. There is no clear difference between these two embryos in the two scenarios. To deal with this issue, a possible solution is to add another camera to monitor the vertical movement of the pipette [21]. However, it increases the system complexity at the same time. Alternatively, the problem is easier to solve by resorting to the force feedback. Specifically, when the pipette needle contacts with target embryo, the force sensor will provide a force signal. Moreover, it is easy to identify the cell state, i.e., whether pierced or unpierced. The force magnitude will drop dramatically once the cell membrane is punctured, which can be detected by the integrated force sensor in the proposed injection system. For illustration, Fig. 4.10 shows the force sensor output signals for the pierced and unpierced cells. The difference between the pierced and unpierced situations can be identified easily from the force signals.

Fig. 4.9 Comparison between different injection scenarios. **a** Micropipette is puncturing the embryo; **b** micropipette slips over the embryo

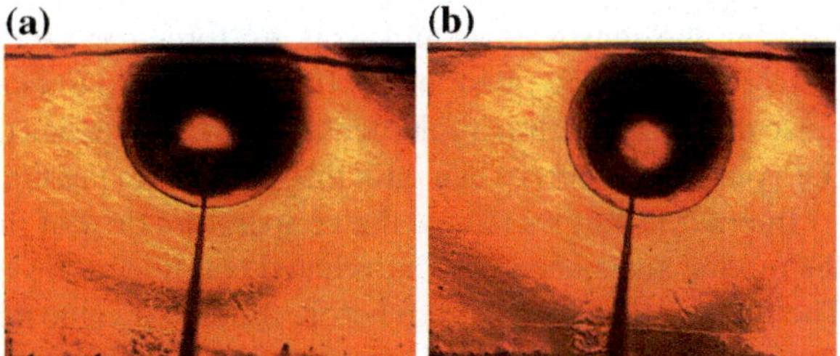

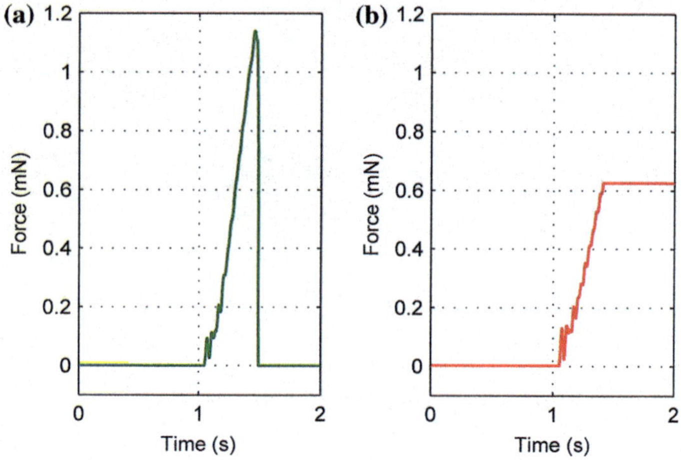

Fig. 4.10 Force sensor output signals during the cell injection. **a** Force signal of pierced cell; **b** force signal of unpierced cell

The collected embryos are spread on the V-grooves and aligned with manual assist carefully. A total of 100 zebrafish embryos are divided into to two equal groups, which are continuously injected using the proposed injection system with position and position/force control, respectively. The average injection time for each embryo is reduced to less than 3 s, which is better than that of a proficient human operator. The injection speed is set as 2 mm/s and 20 mN/s for position and force control, respectively. A gentle retracting speed is set as 220 μm/s. The force trigger threshold from coarse to fine movement is selected as 500 μN, the punctured force velocity trigger is chosen as −10 mN/s. In addition, once the embryo is pierced, a time period of 500 ms is reserved for injecting the desired material. The safe distance of the coarse movement is set as 5 mm. The injected embryos are cultured at room temperature of 24 °C. By trial and error, the parameters of PID position controller are tuned as $K_p = 2 \times 10^{-3}$, $K_i = 6 \times 10^{-4}$, and $K_d = 1.5 \times 10^{-3}$. The parameters of PID force controller are adjusted as $K_p = 6 \times 10^{-3}$, $K_i = 9 \times 10^{-4}$, and $K_d = 1 \times 10^{-3}$.

It is observed that the change rate of the temperature drift is quite low (≤10 mN/s) for the force sensor. In practice, an efficient and quick manipulation behavior (∼20 mN/s) is preferred during the cell injection. Hence, the velocity of variation between the sensor noise and working force can be differentiated conveniently. In this work, a change rate limit of 10 mN/s is used to extract the force signal from the noise during algorithm implementation, which means that only a rapid change signal (i.e., puncturing force) can pass this gate while the noise is resisted.

In order to further reduce the temperature drift of the piezoresistive force sensor, the motion sequence of the automated cell injection system is specifically designed. The force output is set as zero before the signal value reaches the threshold of contact force. The puncturing time is quite short (about 0.1 s), which reduces the drift impact further. The force is reset to zero again after each cell membrane is punctured.

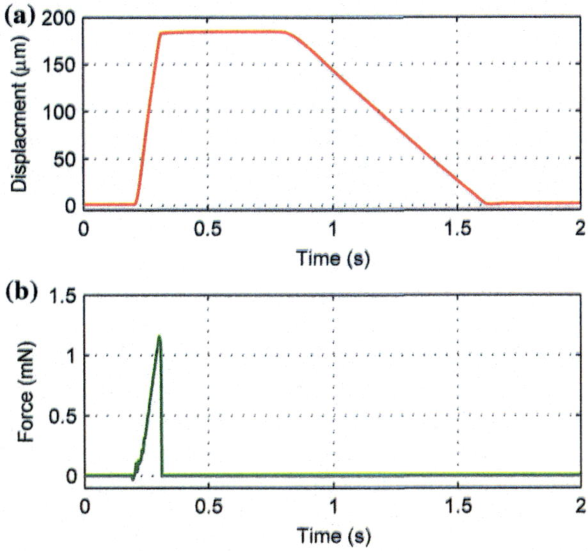

Fig. 4.11 Experimental injection results of one zebrafish embryo with position control. **a** Position trajectory; **b** force trajectory

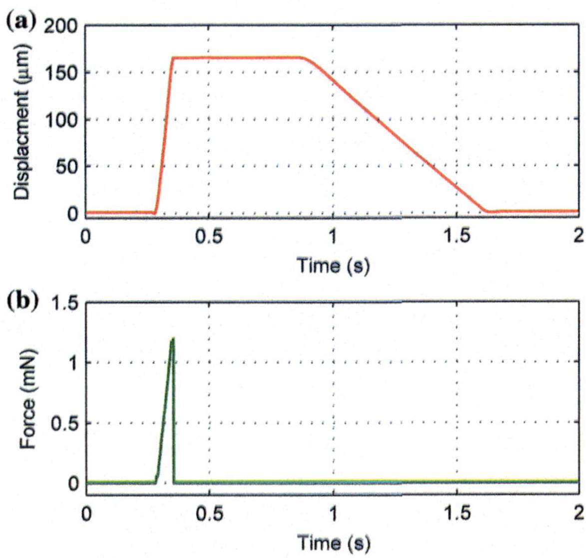

Fig. 4.12 Experimental injection results of one zebrafish embryo with force control. **a** Position trajectory; **b** force trajectory

The experimental results of the position and position/force control are shown in Figs. 4.11 and 4.12, respectively. It is seen that there are some oscillations in the force trajectory when the stand-alone position control is used for cell injection (see Fig. 4.11b), while the force trajectory is smoother with the force control applied (see Fig. 4.12b).

In order to quantitatively evaluate the performance of the automated injection system, the success rate and survival rate are usually adopted as evaluation criteria to characterize a cell injection system [22]. The success rate is defined as the ratio between the number of embryos with a dramatic drop force curve and the total number of injected embryos. With the force feedback, the unsuccessfully injected embryos can be detected and eliminated during the cell injection process. In this work, all of the manipulated cells are injected successfully.

In addition, the survival rate is specified as the ratio between the number of injected embryos which are able to develop into larva and the total number of injected embryos. It essentially represents the severity and frequency of cell damage during the injection. Based on the 100 injected zebrafish embryos with position control and force control, the survival rates are calculated as 82 and 86%, respectively.

The survival rates obtained by the existing approaches are compared in Table 4.2. It is observed that the force control produces a higher survival rate than conventional position control in the same operation environment. With force control involved, the achieved survival rate of 86% is better than majority of existing works. It is notable that the survival rate can be further improved with more delicate care during the embryo culturing and precision control algorithm design. Even so, it is better than the survival rate of 50–70% in manual operation which is caused by the proficiency difference and fatigue problems of human operator. The improvement of the survival rate with force control is dominantly contributed by the smoother force interaction and the greatly reduced embryo deformation during the injection process.

With the force-controlled injection system, another remarkable improvement is the injection speed. With the advantages of force feedback and PZTs, quick response and assured pierced state are achieved without time delay and misjudgment. The injection time of 3 s per embryo can be achieved, i.e., with the injection speed of 20 embryos per minute. The injection speed is much quicker than the manual speed of about 8–15 embryos per minute. In a common practical experiment, the cell number to be processed generally is about 300–500. The improvement of the cell response speed can reduce the individual difference effectively for practical applications, such as drug development and gene therapy.

Table 4.2 Comparison of experimental microinjection results

Method	Cell type (number)	Survival rate (%)
Manual operation	Zebrafish embryo	50–70
Piezodrill system [16]	mouse oocytes (322)	80
Piezo-driven system [31]	Intracytoplasmic sperm (25)	88
Piezo-driven ultrasonic method [8]	Zebrafish embryo (200)	80.7
Piezo-driven with position control (this work)	Zebrafish embryo (50)	82
Piezo-driven with force control (this work)	Zebrafish embryo (50)	86

4.5 Advanced Position and Force Switching Control Design

In this section, an advanced position/force switching control scheme is developed to further smooth the transition between the position and force control process.

4.5.1 *Weight-Based Switching Control System*

Concerning the microinjection procedure of zebrafish embryos, the manual operation can be divided into four stages including approaching, penetrating, injecting, and retracting phases. First, the embryos in a petri dish are identified by then operator, and then, the micropipette is slowly moved closer to the target embryo. When the micropipette tip contacts the chorion of the embryo slightly, the operator drives the micromanipulator manually to provide a rapid thrust movement, which leads to an initial penetration of the chorion by the micropipette tip. Afterward, the embryo is penetrated continuously until the embryo yolk is pierced by the micropipette tip. External materials can be delivered into the embryo at this point by the injector. Then, the micropipette is withdrawn out of the embryo after the injection operation is completed. Nevertheless, the manual operation of microinjection exhibits a low success rate. The reason lies in that it is hard to regulate the position and force simultaneously by a human operator during the microinjection procedure.

To overcome this issue, an automated cell microinjection system can be employed. Specifically, the position control approach is first applied to command the micropipette tip to approach the embryo with a velocity of v_a (μm/s). If the threshold f_{t1} of contact force is attained, the injecting force can be measured with the calibrated force sensor. Meanwhile, force controller instead of position controller is realized to control the injector movement by following a planned force trajectory. Once the membrane is pierced, the position control scheme is adopted again to command the micropipette to finish the remaining operations of the injection process.

Concerning the position/force control scheme, it has one input variable (i.e., the excitation voltage) and two output variables (i.e., the position and force values). The same control input is applied to all of the control phases. Yet, the alternation of different controllers can induce the switching shock and discontinuity of the system input, that can induce a negative effect on the injected cell's viability. To generate a smooth transition between the two different control means, a weight coefficient-based switching method is introduced as shown in Fig. 4.13. In this way, the former controller is fade-out and the latter is fade-in within an overlap of σ time steps. As a result, a smooth transition between the two control processes will be achieved. The control action can be expressed as follows [20].

$$u(k) = \begin{cases} u_p(k)(1-\eta) + u_f(k)\eta & \text{if} \quad t < t_3 \\ u_p(k)\eta + u_f(k)(1-\eta) & \text{if} \quad t \geq t_3 \end{cases} \tag{4.5}$$

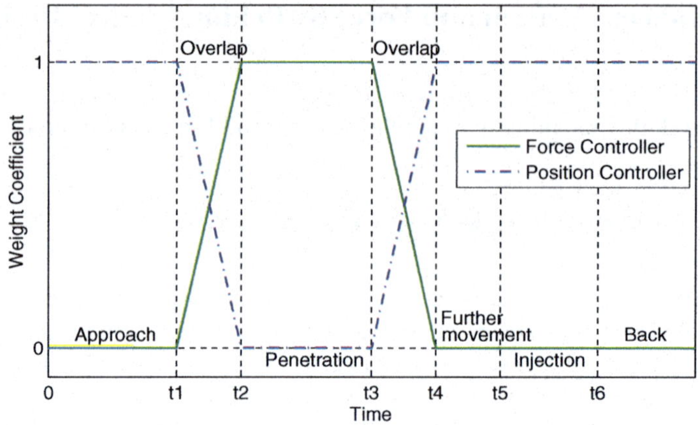

Fig. 4.13 Switching control framework

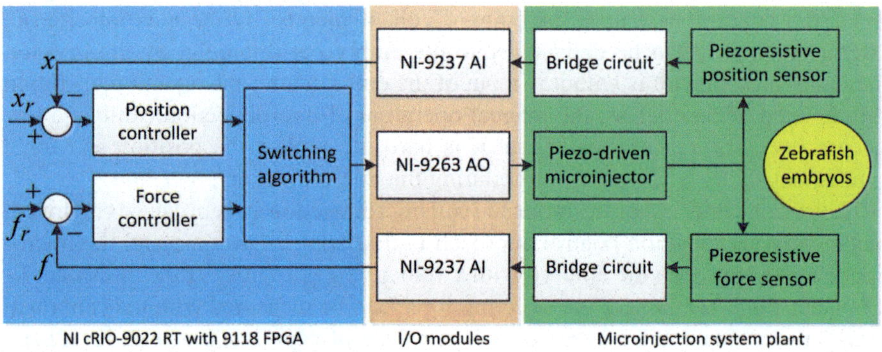

Fig. 4.14 Block diagram of position/force switching control scheme for microinjection

where η represents the weight coefficient, u_p and u_f are the position and force controllers, respectively, and the time t_3 is denoted in Fig. 4.13.

As a model-free control, the incremental PID force controller determines the control action based on its value in the previous time step. In this work, the incremental PID force controller is adopted to limit the change rate of the control signal. In addition, an adaptive sliding mode position controller is implemented to overcome the unpredictable situation during the cell microinjection. Figure 4.14 shows the block diagram of the position/force switching control scheme. In the following sections, the adaptive sliding mode position controller and incremental PID force controller are developed, respectively.

4.5.2 Adaptive Sliding Mode Position Controller Design

To implement a precise position control, the injection system is considered as a mass-spring-damper mechanical system. The whole dynamics model of piezo-driven microinjection system accompanied with uncertain disturbance and nonlinearity (including hysteresis effect) is described by:

$$m\ddot{x}(t) + b\dot{x}(t) + kx(t) = du(t) + h(t) \tag{4.6}$$

where t represents the time variable; x, m, b, and k are the output displacement, mass, damping coefficient, and stiffness for the microinjector, respectively; u denotes the input voltage; d is piezoelectric coefficient; h describes the lumped model disturbance of the system including the hysteresis effect, model parameter uncertainties, and other unknown conditions. In this work, rather than explicitly modeling the hysteresis nonlinearity, the hysteresis effect is considered as a disturbance to the plant model.

To yield the expression for the disturbance $h(t)$, the direct application of the relationship (4.6) results in algebraic loops. Thus, it cannot be implemented practically. In order to solve this problem, the disturbance estimation is conducted on the basis of perturbation estimation technique [4], thanks to its feature of easy to realize.

$$\hat{h}(t) = m\ddot{x}(t) + b\dot{x}(t) + kx(t) - du(t-T) \tag{4.7}$$

where T represents the sampling time interval, and $u(t-T)$ is the control input in the previous time step. In practice, the sampling frequency is selected sufficiently high to make sure that $u(t) = u(t-T)$ approximately.

The employment of the delayed input $u(t-T)$ introduces an estimation error for the lumped disturbance $h(t)$. The estimation error can be expressed as:

$$\tilde{h}(t) = h(t) - \hat{h}(t). \tag{4.8}$$

In view of Eq. (4.8), the system model (4.6) becomes

$$m\ddot{x}(t) + b\dot{x}(t) + kx(t) = du(t) + h(t) + \tilde{h}(t). \tag{4.9}$$

In practice, the input voltage (u) and output position (x) of the piezo-driven microinjection device are limited. Hence, it is reasonable to suppose that the estimation error $\tilde{h}(t)$ is bounded.

Equation (4.7) reveals that the system's full states ($x, \dot{x}, \ddot{x}$) are needed to realize the perturbation estimation. In reality, only the position information x is offered by the displacement sensor. Therefore, in this work, other states ($\dot{x}, \ddot{x}$) are calculated based upon the backward difference equation shown below, thanks to its computational efficiency.

$$x^{(n)}(t) = \frac{x^{(n-1)}(t) - x^{(n-1)}(t-T)}{T} \tag{4.10}$$

We can observe from Eq. (4.6) that all the disturbances are ignored during the system parameter estimation. Nevertheless, the disturbance estimation error $\tilde{h}(t)$ in Eq. (4.8) is presented during the disturbance estimation process. In order to achieve a precise position control by suppressing these disturbances, a scheme of adaptive control is developed in the subsequent discussion to guarantee the robustness of the control system.

Firstly, the output position tracking error is defined as follows.

$$e(t) = x(t) - x_d(t) \tag{4.11}$$

where x and x_d denote the actual and desired position output, respectively.

Based upon the position error (4.11), a PID type of sliding function is defined as

$$s(t) = \dot{e}(t) + \lambda_1 e(t) + \lambda_2 \int_0^t e(\tau) d\tau \tag{4.12}$$

where $\lambda_1 > 0$ and $\lambda_2 > 0$ are positive control parameters.

By recording $\dot{x}_r = \dot{x}_d - \lambda_1 e - \lambda_2 \int_0^t e(\tau) \mathrm{d}\tau$ and ignoring the time variable t, the sliding surface can be rewritten below.

$$s = \dot{x} - \dot{x}_r \tag{4.13}$$

Let

$$\begin{aligned} \mathbf{y} &= [\ddot{x}_r \; \dot{x} \; x] \\ \mathbf{a} &= [m \; b \; k]^{\mathrm{T}} \end{aligned} \tag{4.14}$$

and $\hat{\mathbf{a}}$ describe the estimation for $\mathbf{a}$ with the estimation error

$$\tilde{\mathbf{a}}(t) = \hat{\mathbf{a}}(t) - \mathbf{a}. \tag{4.15}$$

The control action is designed as follows.

$$u = \frac{1}{d}[\mathbf{y}\hat{\mathbf{a}} - k\mathrm{sign}(s) - \hat{h}] \tag{4.16}$$

with the adaptive law

$$\dot{\hat{\mathbf{a}}} = -\mathbf{P}\mathbf{y}^{\mathrm{T}} s \tag{4.17}$$

where $\hat{h}$ represents the estimation of h and $\mathbf{P}$ denotes the design matrix that is symmetric and positive definite.

To demonstrate the stability of the devised controller, a Lyapunov function candidate is chosen as

$$V = \frac{1}{2} m s^2 + \frac{1}{2} \tilde{\mathbf{a}}^{\mathrm{T}} \mathbf{P}^{-1} \tilde{\mathbf{a}} \tag{4.18}$$

The time derivative of V is

$$\dot{V} = sm\dot{s} + \dot{\hat{\mathbf{a}}}^{\mathrm{T}}\mathbf{P}^{-1}\tilde{\mathbf{a}} \tag{4.19}$$

By taking the time derivative of Eq. (4.13) and considering Eq. (4.9), gives

$$\begin{aligned} \dot{s} &= \ddot{x} - \ddot{x}_r \\ &= -\frac{b}{m}\dot{x} - \frac{k}{m}x - \ddot{x}_r + \frac{d}{m}u + \frac{\hat{h} + \tilde{h}}{m} \end{aligned} \tag{4.20}$$

Substituting Eq. (4.20) into (4.19) and taking into account the control law (4.16) and adaptive rule (4.17), results in

$$\begin{aligned} \dot{V} &= s(du + \hat{h} + \tilde{h} - b\dot{x} - kx - m\ddot{x}_r) + \dot{\hat{\mathbf{a}}}^{\mathrm{T}}\mathbf{P}^{-1}\tilde{\mathbf{a}} \\ &= -ks \cdot sign(s) + s\mathbf{y}\tilde{\mathbf{a}} + s\tilde{h} + \dot{\hat{\mathbf{a}}}^{\mathrm{T}}\mathbf{P}^{-1}\tilde{\mathbf{a}} \\ &= -k\,|s| + s\tilde{h} \end{aligned} \tag{4.21}$$

where $s\mathbf{y}\tilde{\mathbf{a}} + \dot{\hat{\mathbf{a}}}^{\mathrm{T}}\mathbf{P}^{-1}\tilde{\mathbf{a}} = 0$ is derived in consideration of the adaptive rule (4.17).

To facilitate the analysis, it is supposed that the estimation error for the lumped disturbance is bounded as follows.

$$|\tilde{h}| \le f \tag{4.22}$$

To ensure that $\dot{V} \le 0$, the control gain is selected according to:

$$k \ge \eta + f \tag{4.23}$$

where the constant η is positive.

Then, in view of (4.21), $\dot{V}$ meets the condition:

$$\dot{V} \le -\eta|s| \tag{4.24}$$

which indicates that $\dot{V}$ is negative definite.

In addition, the derivative of $\dot{V}$ is calculated by

$$\ddot{V} = \begin{cases} -k\dot{s} + \dot{s}\tilde{h}, & \text{if } s > 0 \\ k\dot{s} + \dot{s}\tilde{h}, & \text{if } s \le 0 \end{cases} \tag{4.25}$$

where k is a positive constant. Then, we can derive that $\ddot{V}$ is bounded. In accordance with the definition of (4.12) and (4.17), we can deduce that the sliding function variable $s \to 0$ as $t \to +\infty$. It follows that the tracking error meets $\lim\limits_{t \to +\infty} e(t) = 0$,

$\lim_{t \to +\infty} \dot{e}(t) = 0$ and $x \to x_d$, $\dot{x} \to \dot{x}_d$ as $t \to +\infty$. Hence, the proposed adaptive sliding mode control strategy is stable.

To suppress the chattering effect, the signum function in (4.16) is replaced by a saturation function. That is,

$$\text{sat}(s/\phi) = \begin{cases} \text{sign}(s), & \text{if } |s| > \phi \\ s/\phi, & \text{if } |s| \leq \phi \end{cases} \tag{4.26}$$

where ϕ is the boundary layer thickness. In practice, the parameter ϕ is selected by taking a comprise between the tracking accuracy and chattering effect.

4.5.3 Incremental PID Force Controller Design

During the force control phase, the injection force (f) follows the desired trajectory of force (f_r) by using an incremental PID controller. By defining $e_f(t) = f_r(t) - f(t)$ as the force tracking error, the incremental PID control action is expressed below.

$$\begin{aligned} u_f(t) =& u(t-T) + K_p[e_f(t) - e_f(t-T)] + K_i e_f(t) \\ &+ K_d[e_f(t) - 2e_f(t-T) + e_f(t-2T)] \end{aligned} \tag{4.27}$$

where f represents actual injection force, f_r describes the desired force trajectory, T denotes the sampling time interval, and $u(t-T)$ is the control action in the previous time step. K_p, K_i and K_d are the positive control gains.

In general, the three control gains of PID controller can be adjusted by Ziegler-Nichol method through simulation study, and then finely tuned by experimental study. Nevertheless, owing to the lack of an accurate force model of the cell deformation, experimental studies are directly adopted to tune the PID gains through the trial-and-error method.

4.5.4 Switching Scheme Design

In the approaching phase, firstly, the micropipette is translated from its initial position close to the embryos with a velocity of v_a (μm/s) by position control. If the threshold force f_{t1} is exceeded by the contact force, the position controller is turned off and the incremental PID force controller is turned on within σ time steps. The purpose of the force control is to command the injection force to track the desired trajectory of force f_r (mN) precisely. Once the cell membrane is pierced, the contact force drastically drops, which can be detected by the microforce sensor. When the magnitude of the penetration force reduces down to the threshold ε, the force sensor is switched off and position controller is switched on within σ time steps. In this way, the injector is

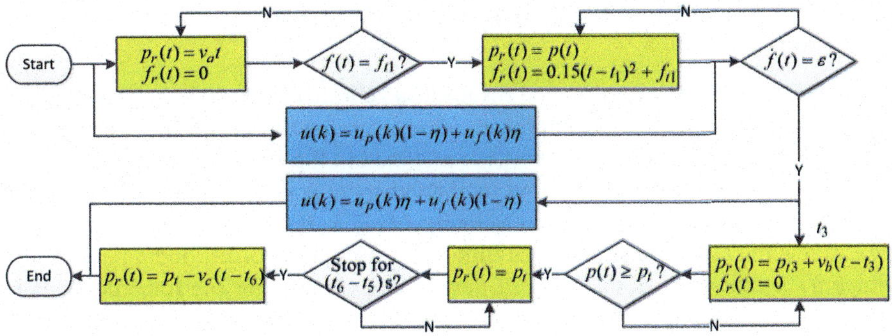

Fig. 4.15 Block diagram of position/force switching control program

commanded by position control to penetrate continuously with a velocity v_b (μm/s) until the end-effector attains the desired position p_t, i.e., the embryo yolk. After that, the pipette tip stops for $(t_6 - t_5)$ seconds to simulate the genetic material delivery. Finally, the micropipette is extracted out of the embryo with a velocity v_c (μm/s). Figure 4.15 depicts the switching process between the position and force controllers.

4.6 Experimental Testing Results

To illustrate the efficiency of the developed injection control scheme with position/force switching algorithm, experimental study of microinjection with zebrafish embryos is conducted in this section.

4.6.1 Controller Setup

For the microinjection control system, the position controller is implemented by the adaptive sliding mode control algorithm, and the force controller is realized by the incremental PID algorithm. The control parameters of the adaptive sliding mode position controller are selected as: $\lambda_1 = 90$ and $\lambda_2 = 0.2$. In addition, the initial value of the mass, damping coefficient, and stiffness are chosen as $m = 1$, $b = 1$, and $k = 1$, respectively. The tuned gains of the incremental PID force controller are: $K_p = 0.06$, $T_i = 0.02$, and $T_d = 0.001$. Besides, the force threshold for the control switching is set as $f_{t1} = 0.05$ mN, the injection time is 1 s at the maximum injection position p_t = 160 μm, and the transition time σ is selected as 0.04 s.

4.6.2 Position/Force Switching Control Results

Figure 4.16 depicts the experimental results of a zebrafish embryo injection with the developed control scheme. The references of position and force trajectories are shown in Figs. 4.16a, b, and the control errors are assigned in Fig. 4.16c, d, respectively. We can observe that the injector tip is commanded by adaptive sliding mode position controller to translate at a constant velocity of $v_a = 100\,\mu\text{m/s}$ until the injecting force attains the force threshold f_{t1} at time $t_1 = 1.13$ s. The corresponding excitation voltage u_1 is extracted as the basis of the control variable for the force controller. After that, the position control is gradually replaced by the incremental PID force control in the time interval σ.

For imitating the manual operation of a skilled human, the injection force of an expert can be recorded and fitted as a quadratic curve [24]. In this research, the following quadratic curve is chosen as the reference force trajectory.

$$f_r(t) = 0.15(t - t_1)^2 + f_{t1} \tag{4.28}$$

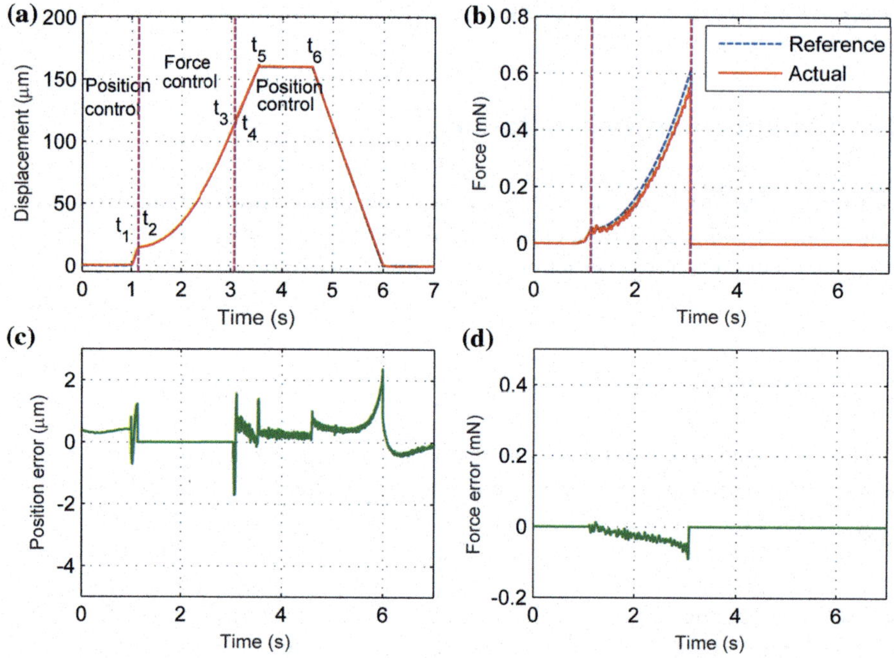

Fig. 4.16 Experimental results of the proposed position/force switching control for zebrafish embryo injection. **a** Position control result; **b** penetration force control result; **c** position control error; **d** force control error

The cell membrane is pierced at the time of $t_3 = 3.07$ s. Afterward, the contact force between the injector and cell drops dramatically, that is detected by the microforce sensor easily. The position control is gradually turned on again within σ time interval, and the microinjector continuously moves at a feed velocity of $v_b = 100\,\mu$m/s. Once the tip of the microinjector arrives at the desired position $p_t = 160\,\mu$m at $t_5 = 3.59$ s, it takes about 1 s to deliver the genetic material into the cell. At last, the injector micropipette is extracted out of the target cell and moves back to the initial position with a velocity of $v_c = 115\,\mu$m/s by the position control after $t_6 = 4.59$ s. Figure 4.18 shows the photo snapshots of the microinjection procedure for a zebrafish embryo.

In order to quantify how well the switching problem is tackled, the same control process has been realized without adopting the introduced switching algorithm for the purpose of comparison. Figure 4.17 shows the experimental results. We can see that there is a position jerk at the moment when the force controller is switched on. A clear force shock is also observed. The peak values of position and force control errors are $-13.56\,\mu$m and 0.71 mN, respectively, which are worse than the position/force switching control results.

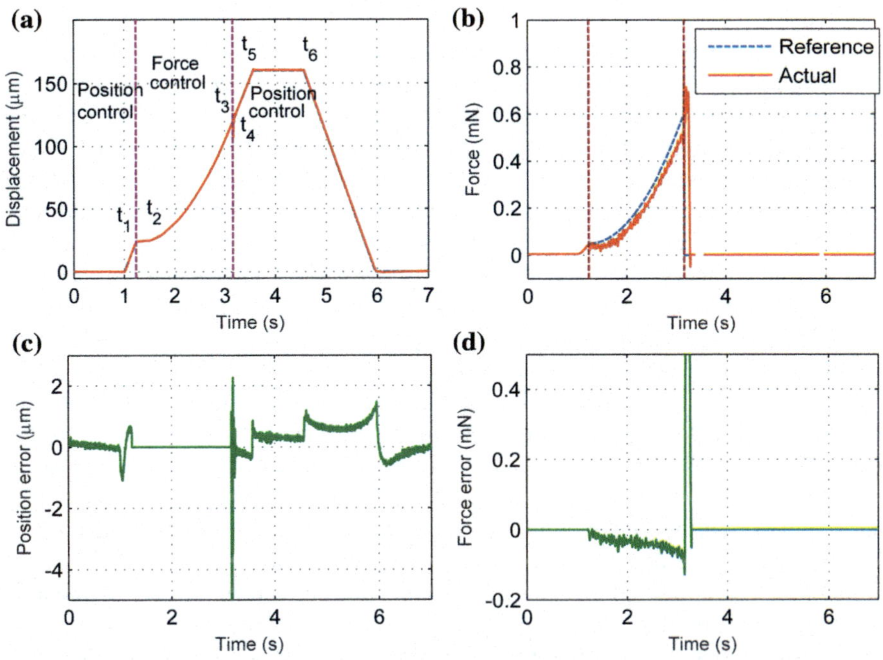

Fig. 4.17 Experimental results of conventional position/force switching control for zebrafish embryo injection. **a** Position control result; **b** penetration force control result; **c** position control error; **d** force control error

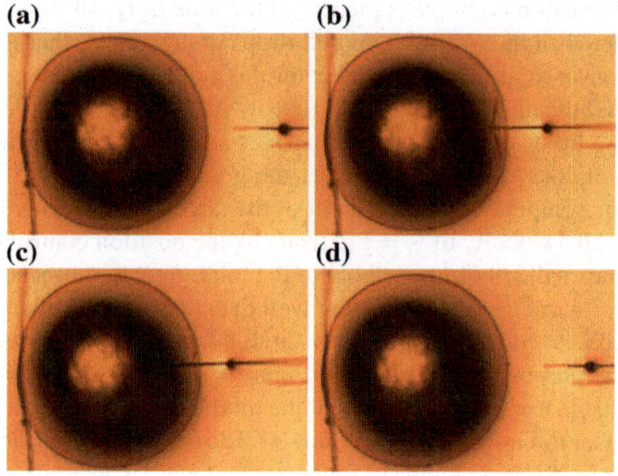

Fig. 4.18 Photos of injection phases for the zebrafish embryo. **a** Approaching phase; **b** penetration phase; **c** injecting phase; **d** retracting phase

4.6.3 Discussions

From the foregoing experimental studies, we can observe from Fig. 4.16a, b that the performance of the proposed system caters for the requirement of precision position/force control with a smooth transition. The implementation of the adaptive sliding mode position controller does not need prior knowledge on the bound values for the uncertainties and unknown parameters of the system. Figure 4.16c reveals that the position tracking error of smaller than 1.72 μm is resulted when the controllers are switched. The largest tracking error of about 2.36 μm is produced by the position control, which occurs at the start and end moments of the micropipette tip's translation. The reason lies is that the parameters of sliding mode position controller are selected to generate a rapid response for the position tracking; hence, the overshoot is induced when the motion state changes.

During the penetration phase, the force control is implemented by PID control algorithm. Its control objective is to offer the desired contact force trajectory in the penetrating stage. The PID control method is adopted because it is hard to establish an accurate dynamics model to characterize the relationship between the contact force and cell's deformation. On the contrary, PID control is a model-free control approach, that does not require an analytical model of the control plant. From Fig. 4.16d, we can observe that no abrupt changes in position and force emerge at the moment of switching between the position and force controllers. The maximum tracking error of 0.09 mN occurs at 3.07 s for the force control, which is much lower than that produced by conventional switching control.

The efficiency of the proposed microinjection system has been validated by experimental study on zebrafish embryo penetration. Experimental results demonstrate that the introduced position/force switching control strategy is feasible for cell microinjection tasks. To improve the applicability of the microinjection system, a cell holding device needs to be designed in the future work. As the target cell is not immobilized firmly in the V-groove in the reverse direction of injection, the penetrated zebrafish embryo will follow the pipette a short distance during the extraction of the micropipette. Promising solutions involve the use of microfluidic devices or fixing plates with multiple walls, which are expected to trap and release embryos with the help of pressure change in through-holes.

4.7 Conclusion

This chapter has reported the design, analysis, and experimental verification of a piezo-driven cell injection system with force feedback. The proposed cell injection system features convenient installation, low cost, and easy maintenance. A flexure-based displacement amplifier is designed to enlarge the stroke of PZT. With the integration of piezoresistive sensors, the position and force status of the injected embryos is monitored in real time. This system enables a high operation speed, high success rate, and high survival rate. Experimental results confirm the superiority of the force control versus conventional position control in the cell injection system. Moreover, the issue of transition jerk at the moment of switching between the force and position controllers is overcome by introducing a weight coefficient method. The proposed microinjection system can be easily applied to other cell injection applications, such as mouse oocytes, graine, and other types of suspended biological cells.

References

1. Ammi, M., Ferreira, A.: Biological cell injection visual and haptic interface. Adv. Robot. **20**(3), 283–304 (2006)
2. Avci, E., Hattori, T., Kamiyama, K., Kojima, M., Horade, M., Mae, Y., Arai, T.: Piezo-actuated parallel mechanism for biological cell release at high speed. Biomed. Microdevices **17**(5), 98 (2015)
3. Ediz, K., Olgac, N.: Effect of mercury column on the microdynamics of the piezo-driven pipettes. J. Biomech. Eng. **127**(3), 531–535 (2005)
4. Elmali, H., Olgac, N.: Implementation of sliding mode control with perturbation estimation (SMCPE). IEEE Trans. Control Syst. Technol. **4**(1), 79–85 (1996)
5. Fan, M., Feng, Z., Agca, Y., Critser, J.K.: Vibration study of the piezodriven pipettes immersed in viscous liquids. J. Appl. Phys. **100**(7), 074701 (2006)
6. Gan, Y., Chen, Z.: A study of the zona piercing process in piezodriven intracytoplasmic sperm injection. J. Appl. Phys. **104**(4), 044702 (2008)
7. Howe, K., Clark, M.D., Torroja, C.F., Torrance, J., Berthelot, C., Muffato, M., Collins, J.E., Humphray, S., McLaren, K., Matthews, L., et al.: The zebrafish reference genome sequence and its relationship to the human genome. Nature **496**(7446), 498–503 (2013)
8. Huang, H., Mills, J.K., Lu, C., Sun, D.: A universal piezo-driven ultrasonic cell microinjection system. Biomed. Microdevices **13**(4), 743–752 (2011)
9. Huang, H.B., Sun, D., Mills, J.K., Cheng, S.H.: Robotic cell injection system with position and force control: toward automatic batch biomanipulation. IEEE Trans. Robot. **25**(3), 727–737 (2009)

10. Huang, T., Kimura, Y., Yanagimachi, R.: The use of piezo micromanipulation for intracytoplasmic sperm injection of human oocytes. J. Assist. Reprod. Genet. **13**(4), 320–328 (1996)
11. Karimirad, F., Chauhan, S., Shirinzadeh, B.: Vision-based force measurement using neural networks for biological cell microinjection. J. Biomech. **47**(5), 1157–1163 (2014)
12. Liu, Y., Zhang, X., Zhang, Y., Xu, Q.: Design and control of a novel piezo-driven micro-injector. In: Proceedings of 2015 IEEE 10th International Conference on Robotics and Biomimetics, pp. 1985–1990. IEEE (2015)
13. Lu, Z., Chen, P.C., Nam, J., Ge, R., Lin, W.: A micromanipulation system with dynamic force-feedback for automatic batch microinjection. J. Micromech. Microeng. **17**(2), 314 (2007)
14. Permana, S., Grant, E., Walker, G.M., Yoder, J.A.: A review of automated microinjection systems for single cells in the embryogenesis stage. IEEE/ASME Trans. Mech. **21**(5), 2391–2404 (2016)
15. Pillarisetti, A., Pekarev, M., Brooks, A.D., Desai, J.P.: Evaluating the effect of force feedback in cell injection. IEEE Trans. Autom. Sci. Eng. **4**(3), 322–331 (2007)
16. Ron-El, R., Liu, J., Nagy, Z., Joris, H., Van den Abbeel, E., Van Steirteghem, A.: Intracytoplasmic sperm injection in the mouse. Hum. Reprod. **10**(11), 2831–2834 (1995)
17. Shen, Y., Fukuda, T.: State of the art: micro-nanorobotic manipulation in single cell analysis. Robot. Biomimetics **1**(1), 21 (2014)
18. Sun, Y., Nelson, B.J.: Biological cell injection using an autonomous microrobotic system. Int. J. Robot. Res. **21**(10–11), 861–868 (2002)
19. Wang, G., Xu, Q.: Design and development of a piezo-driven microinjection system with force feedback. Adv. Robot. **31**, 23–24 (2017). https://doi.org/10.1080/01691864.2017.1362996
20. Wang, G., Xu, Q.: Design and precision position/force control of a piezo-driven microinjection system. IEEE/ASME Trans. Mechatron. **22**(4), 1744–1754 (2017)
21. Wang, W., Liu, X., Sun, Y.: Contact detection in microrobotic manipulation. Int. J. Robot. Res. **26**(8), 821–828 (2007)
22. Wang, W.H., Liu, X.Y., Sun, Y., et al.: High-throughput automated injection of individual biological cells. IEEE Trans. Autom. Sci. Eng. **6**(2), 209–219 (2009)
23. Wei, Y., Xu, Q.: Design of a PVDF-MFC force sensor for robot-assisted single cell injection. IEEE Sens. J. **17**(13), 3975–3982 (2017)
24. Xie, Y., Sun, D., Liu, C., Tse, H.Y., Cheng, S.H.: A force control approach to a robot-assisted cell microinjection system. Int. J. Robot. Res. **29**(9), 1222–1232 (2010)
25. Xie, Y., Sun, D., Tse, H.Y.G., Liu, C., Cheng, S.H.: Force sensing and manipulation strategy in robot-assisted microinjection on zebrafish embryos. IEEE/ASME Trans. Mechatron. **16**(6), 1002–1010 (2011)
26. Xu, Q.: Design and smooth position/force switching control of a miniature gripper for automated microhandling. IEEE Trans. Ind. Inform. **10**(2), 1023–1032 (2014)
27. Xu, Q.: Digital integral terminal sliding mode predictive control of piezoelectric-driven motion system. IEEE Trans. Ind. Electron. **63**(6), 3976–3984 (2016)
28. Xu, Q.: Precision motion control of piezoelectric nanopositioning stage with chattering-free adaptive sliding mode control. IEEE Trans. Autom. Sci. Eng. **14**(1), 238–248 (2017)
29. Xu, Q., Li, Y.: Analytical modeling, optimization and testing of a compound bridge-type compliant displacement amplifier. Mech. Mach. Theory **46**(2), 183–200 (2011)
30. Yan, X., Sun, D.: Multilevel-based topology design and cell patterning with robotically controlled optical tweezers. IEEE Trans. Control Syst. Technol. **23**(1), 176–185 (2015)
31. Yi, Z., KokKiong, T., Sunan, H.: Software based vision system for automated cell injection. In: Proceedings of International Conference on BioMedical Engineering and Informatics (BMEI 2008), vol. 1, pp. 718–722. IEEE (2008)
32. Yu, Y., Ding, C., Wang, E., Chen, X., Li, X., Zhao, C., Fan, Y., Wang, L., Beaujean, N., Zhou, Q.: Piezo-assisted nuclear transfer affects cloning efficiency and may cause apoptosis. Reproduction **133**, 947–954 (2007)
33. Zhang, Y., Tan, K.K., Huang, S.: Vision-servo system for automated cell injection. IEEE Trans. Ind. Electron. **56**(1), 231–238 (2009)

Chapter 5
Design, Fabrication, and Testing of a Constant-Force Microinjector

Abstract This chapter presents the design and testing of a flexure-based microinjector with constant force output dedicated to biological cell micromanipulation. The microinjector offers a constant force without adopting a force controller. The motion control is sufficient to provide a constant output force, that simplifies the system design procedure. The injector is actuated by a piezoelectric actuator via a displacement amplifier. Analytical models of the mechanism are established and verified by conducting simulation study with finite element analysis (FEA). A prototype device is fabricated by 3D printing process for experimental study. The feasibility of the developed constant-force injector for biological cell micromanipulation is verified by experimental studies.

5.1 Introduction

Compliant mechanism works on the basis of the deformation of its internal flexible members [11]. In the literature, lots of flexure-based compliant positioning mechanisms have been devised for micro-/nanopositioning applications [3, 4, 15]. In recent years, the requirement on constant force emerges in the applications of micropositioning stages. Conventionally, the constant force output is achieved by adopting a process control [22, 23, 25]. The controller is usually realized by a switching between the displacement and force control, that complicates the control process and degrades the control accuracy. In order to overcome the contradiction between the position and force control accuracy, it is necessary to realize the constant force output for the positioning mechanisms by mechanical design approach [1, 7].

In comparison with traditional variable-force mechanism, constant-force mechanism exhibits several remarkable advantages. Firstly, a constant-force mechanism allows the reduction of the driving force, that can extend the output displacement indirectly. Secondly, it allows the elimination of use of a force controller and hence decreases cost on hardware. In the literature, some promising applications of constant-force mechanisms have been reported, including constant-force micropositioning stage [21], caster wheel [5], grippers [10, 13, 19], and others [2, 9, 14].

Q. Xu, *Micromachines for Biological Micromanipulation*,
https://doi.org/10.1007/978-3-319-74621-0_5

This chapter introduces the design of a flexure-based compliant constant-force mechanism dedicated to biological cell micromanipulation. To achieve a constant force, the idea of zero-stiffness design is realized by combining a positive-stiffness and a negative-stiffness mechanisms. In addition, the modified leaf flexures (MLFs) are introduced as the positive-stiffness mechanism to generate a compact design and to eliminate the stress stiffening phenomenon. Bistable beams are adopted as the negative-stiffness mechanism [18]. Unlike existing constant-force mechanisms that are composed of curved beams [19], the presented one is constructed by standard beams. As a result, the reported constant-force mechanism offers a simpler structure and enables easier fabrication.

Concerning the actuation approach, piezoelectric actuator (PZT) is a popular choice as it has a number of advantages, such as high motion resolution, high response speed, high power density, and compact structure [6, 12, 16]. Therefore, the proposed constant-force mechanism is actuated by a PZT via a displacement amplifier [20]. Parametric design is carried out to offer an output displacement of over 100 μm. The mechanism performance is verified by performing finite element analysis (FEA) simulations. Moreover, the stage is employed as a cell microinjector. A prototype has been fabricated using a 3D printer for experimental tests. To illustrate its applications in biological manipulation, several experimental studies have been conducted using crab eggs.

5.2 Structure Design

The structural design of a precision positioning stage with constant force output is presented in this section. It includes the contents of displacement amplifier design, zero-stiffness structure design, parameter optimization, and layout design.

5.2.1 Design of Displacement Amplifier

Although PZT has been extensively adopted in ultrahigh-precision positioning devices, PZT actuator exhibits a critical disadvantage of limited stroke output [24]. For practical applications, a displacement amplifier is required to magnify the output stroke of the PZT. Currently, different approaches are available to magnify the PZT's displacement, e.g., leverage-principle amplifier, instable-principle lever amplifier. Generally, the amplifiers based on the triangular principle exhibit the advantages of high efficiency, simple mechanism, and compact structure. Thus, a bridge type of amplifier is utilized as depicted in Fig. 5.1a, which is adopted to amplify the output stroke for PZT actuator.

In particular, the bridge type of amplification mechanism is employed to amplify the displacement on the basis of triangular principle. The main parameters are described in Fig. 5.1b. The schematic diagram for the working principle is illustrated

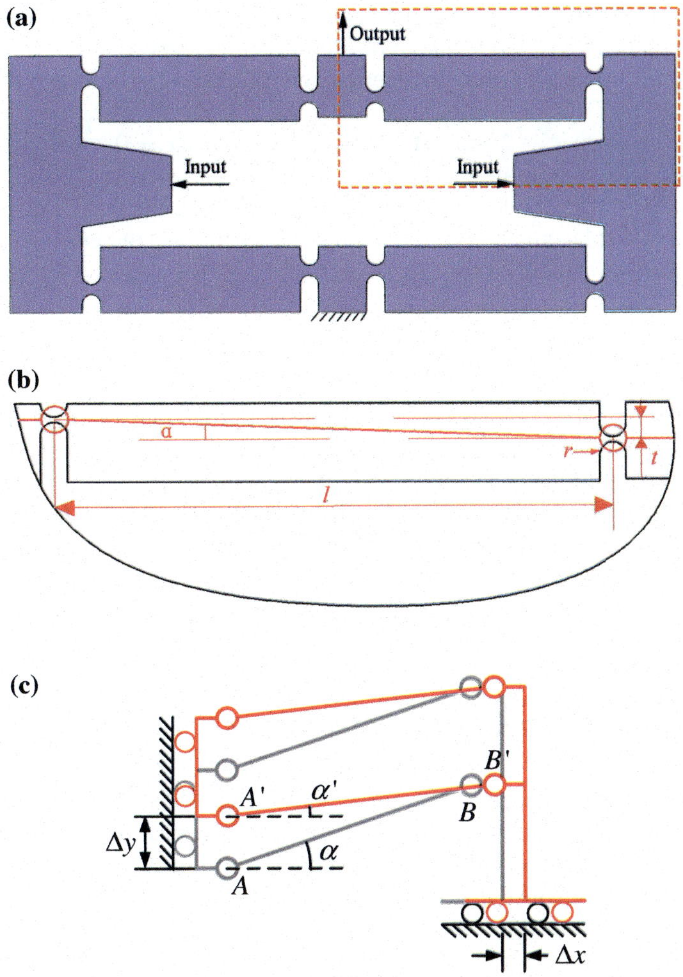

Fig. 5.1 Illustration of **a** bridge-type displacement amplifier, **b** main parameters, and **c** one-quarter (upper-right corner) model of the displacement amplifier

in Fig. 5.1c, where Δy and Δx denote the output displacement and input displacement, respectively. An increase of input displacement Δx induces the reduction of the angle α, which leads to an increase of output displacement Δy. In view of the relationships among the three sides for a right triangle, we can obtain the following equations.

$$l \cos \alpha' = l \cos \alpha + \Delta x \tag{5.1}$$

$$l \sin \alpha' = l \sin \alpha - \Delta y \tag{5.2}$$

where l represents the length for the segment AB. In addition, α and α' describe the initial angle and changed angle due to the deformation, respectively.

By using the property of triangle function, the variable α can be eliminated, which results in:

$$\Delta y^2 - 2l \sin\alpha \Delta y + \Delta x^2 + 2l \cos\alpha \Delta x = 0 \tag{5.3}$$

In comparison with the length l, the squared values for the output displacement Δy^2 and input displacement Δx^2 are negligible. Therefore, the expression for the amplification ratio is reduced to:

$$A = \frac{\Delta y}{\Delta x} = \frac{\cos\alpha}{\sin\alpha} = \cot\alpha \tag{5.4}$$

After generating the model of the bridge-type amplifier, the dominant issue of the design is the appropriate selection of the flexure hinge. We assume that the flexure hinge functions as a 1-DOF rotational joint in the analytical model. To yield a good performance, two characteristics (i.e., small input stiffness and large output stiffness) are expected for the displacement amplifier. Specifically, as for the bridge type of amplifier, it should exhibit small rotational stiffness and large translational stiffness. To cater for such requirements, the right-circular hinge is adopted in the case design.

5.2.2 Design of Zero-Stiffness Mechanism

In this work, the zero-stiffness mechanism is constructed by the parallel connection of a negative-stiffness mechanism and a positive-stiffness mechanism. Thus, the design process of the zero-stiffness mechanism includes the design of negative-stiffness and positive-stiffness mechanisms, respectively.

5.2.2.1 Design of Negative-Stiffness Mechanism

Firstly, the design of negative-stiffness mechanism is performed. A fixed-guided straight beam exhibits a bistable behavior when it is arranged tiltedly. Under large deformation, the tilted beam experiences buckling phenomenon with negative stiffness. In this case design, the negative-stiffness mechanism is constructed by bistable beam which experiences fixed-guided constraint condition.

Before the use of the mechanism, it is necessary to predict its performance reliably. In the literature, a number of approaches have been presented to evaluate the performance of bistable beams. These include the pseudo-rigid-body model (PRBM) [8], elliptic integral model [18], and finite element model [17]. Each approach possesses specific advantages and disadvantages. In this work, the elliptic integral model is adopted to design the parameters for the mechanism, which is then verified by employing finite element model.

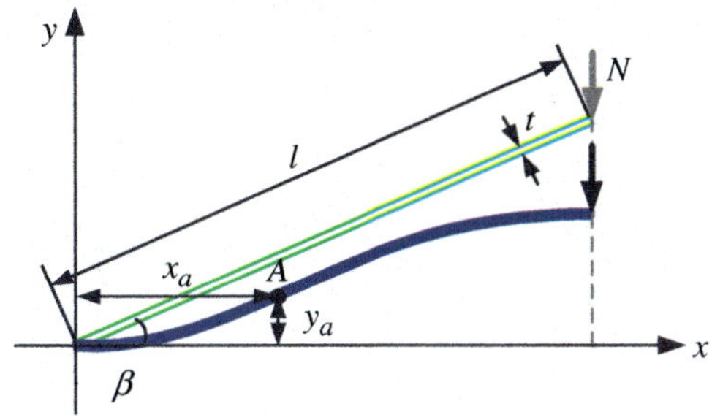

Fig. 5.2 Schematic diagram of a bistable beam with parameters

Figure 5.2 depicts the schematic diagram for a bistable beam along with force conditions. The procedures of analytical modeling have been reported in the literature, e.g., [21]. The dominant analytical models are presented below.

$$\begin{aligned}\frac{y_a}{L} &= -\frac{1}{R}\{2\sin\beta[E(u,\alpha_2) - E(u,\alpha_1) - F(u,\alpha_2) \\ &\quad + F(u,\alpha_1)] + 2u\cos\beta(\cos\alpha_1 - \cos\alpha_2)\}\end{aligned} \tag{5.5}$$

$$\begin{aligned}\frac{x_a}{L} &= -\frac{1}{R}\{2\cos\beta[E(u,\alpha_2) - E(u,\alpha_1) - F(u,\alpha_2) \\ &\quad + F(u,\alpha_1)] + 2u\cos\beta(\cos\alpha_2 - \cos\alpha_1)\}\end{aligned} \tag{5.6}$$

where x_a and y_a denote the coordinates for the point A, β is the initial inclined angle of the beam, R represents the reaction force of the beam, and α describes the variable of the beam after deformation. In addition, $E(\cdot)$ and $F(\cdot)$ are the first and second kind of elliptic integral, and u represents a defined value. More details can be found in the literature [21].

By utilizing the analytical model, we can predict the force–displacement behavior of the bistable beam conveniently.

5.2.2.2 Design of Positive-Stiffness Mechanism

Previously, the leaf flexure with slender shape is usually adopted in the design of positive-stiffness mechanism. Nevertheless, the conventional leaf flexure exhibits two disadvantages. First, a slender leaf flexure commonly possesses a large physical size. In consideration of the limit on fabrication capacity, a long rather than a thin leaf flexure is usually adopted. It leads to a larger physical size for the entire mechanism. In addition, it causes an adverse effect on the loading capacity. Second, the conventional leaf flexure usually exhibits stress stiffening effect. In particular, evident stress stiffening can be experienced by a slender leaf flexure due to the motion of the mechanism. Under the effect of compressive stress, stress stiffening results in the increase of transverse stiffness for the leaf flexure.

In order to enhance the performance of the mechanism, the modified leaf flexure (MLF) is adopted as the positive-stiffness mechanism in this work. Figure 5.3

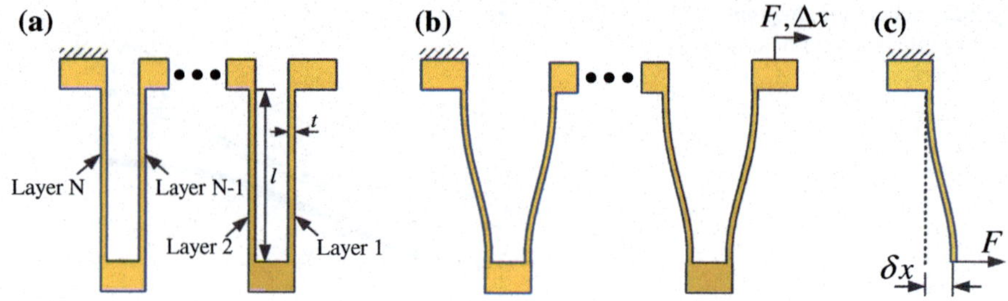

Fig. 5.3 Sketch diagrams of the modified leaf flexure (MLF). **a** Initial state; **b** deformed shape; **c** deformation of one leaf flexure

illustrates the idea of MLF with N modules. The output end is connected to the base via N modules of leaf flexure. All of the flexures of MLF have the identical length l. If an external force F is exerted on the output end, the MLF experiences a deflection. Each flexure suffers from a combined force and moment. Based on the knowledge on mechanics of materials, we can obtain the following relations:

$$\frac{Fl^2}{2EI} - \frac{Ml}{EI} = 0 \tag{5.7}$$

$$\frac{Fl^3}{3EI} - \frac{Ml^2}{2EI} = \delta x \tag{5.8}$$

Substituting Eq. (5.7) into (5.8) yields

$$\delta x = \frac{Fl^3}{12EI} \tag{5.9}$$

The spring constant of the MLF is calculated by

$$k = \frac{F_0}{\Delta x} = \frac{F}{N\delta x} = \frac{Ebh^3}{Nl^3} \tag{5.10}$$

From Eq. (5.10), we can deduce that under the identical length of leaf flexures, the stiffness of MLF is $1/N$ of a leaf flexure. Besides, owing to the avoidance of overconstraints, the stress stiffening effect is removed. Equation (5.10) can be adopted to design the parameters for the MLF.

5.2.3 *Parametric Study*

The stage's performance is dominantly affected by its parametric design. Hence, a parametric study is carried out to provide the guideline for determining optimal parameters of the stage. The stage consists of both rigid and flexible elements. While the rigid elements are used to support the stage structure, the flexible elements are mainly responsible for the stage's output performance. Unlike the slender leaf flexures in the positive-stiffness mechanism, the bistable beams exhibit buckling behavior which is more sensitive to the designed parameters. Furthermore, regarding the design order, the parameters for the bistable beams are firstly designed. And then, the parameters of MLF are designed in accordance with the performance of the designed bistable beams. Thus, the influences of parametric design for the bistable beam are dominantly discussed in the following. In particular, bistable beams have four main parameters including the inclination angle, length, out-of-plane thickness, and in-plane width. For a parametric study, each parameter is changed and the influence on the performance of bistable beam is obtained by resorting to the analytical model.

Firstly, the inclination angle for the bistable beam is increased gradually from 3.5° to 4.5° with a step size of 0.1°. Figure 5.4a depicts the obtained force–displacement relationships. It is observed that the force magnitude arises as the inclination angle increases. An increase of 0.1° for the inclination angle causes an increase of about 0.02 N for the force. The relationship between the force and inclination angle is not directly proportional. Additionally, as the inclination angle increases, the constant-force stroke increases as well. The results demonstrate the variation of the force–displacement behavior for the bistable beam around the inclination angle of 4°.

Secondly, the length of the bistable beam is gradually changed from 30 to 39 mm with a step size of 1 mm. The results of force–deflection behavior for the bistable beam are depicted in Fig. 5.4b. We can find that as the length increases, the force magnitude decreases. Moreover, we can observe that the shorter the beam length, the larger the length's influence on the force magnitude. In addition, as the length increases, the bistable beam is easier to experience the buckling behavior.

Thirdly, the out-of-plane thickness of the bistable beam is varied between 2.6 and 3.4 mm with an interval of 0.2 mm. The force–deflection behaviors of the bistable beam are obtained as illustrated in Fig. 5.4c. We can see that as the thickness increases, the force magnitude increases as well. The relationship between the force and out-of-plane thickness is approximately of direct proportional. Besides, the variation of the out-of-plane thickness imposes nearly no effect on the constant-force stroke.

Finally, the in-plane width of the bistable beam is increased gradually from 0.5 to 1.3 mm with 0.2-mm step size. The generated force–deflection relations are displayed in Fig. 5.4d. In comparison with the other three parameters, the in-plane width imposes the greatest influence on the force output. Specifically, the force increases as the width arises. In addition, the larger the beam width, the greater the influence on the force magnitude.

The foregoing analysis indicates that the in-plane width of the bistable beam is the most influential design parameter for the performance of the bistable beam. At the same time, from the viewpoint of fabrication process, the designed inclination angle of the bistable beam is more difficult to be ensured.

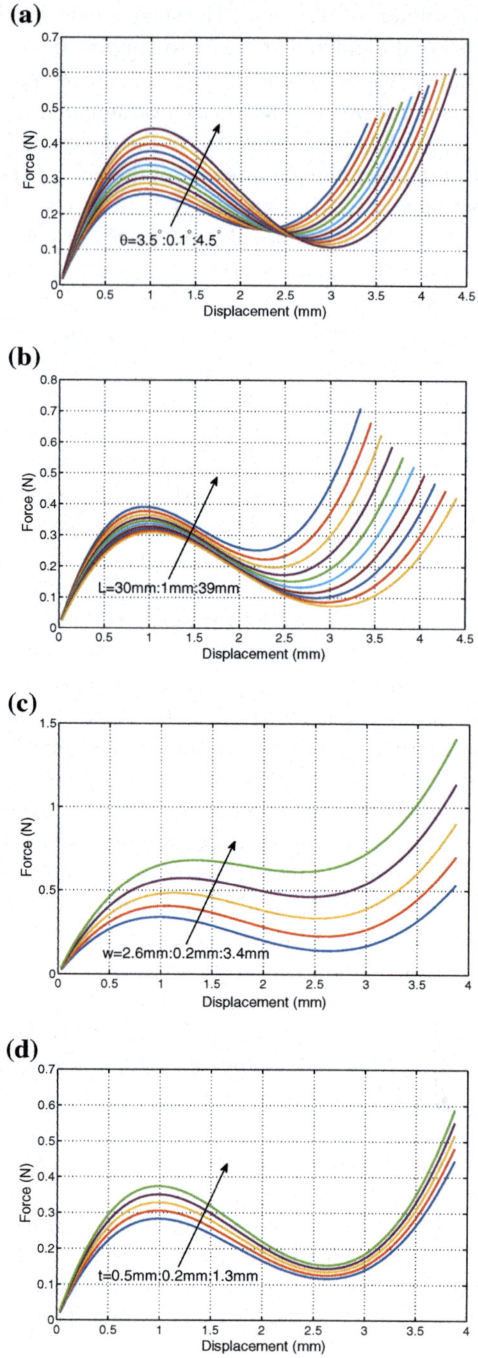

Fig. 5.4 Simulation results of the force–displacement relationship when changing **a** inclination angle, **b** length, **c** out-of-plane thickness, and **d** in-plane width of the bistable beams

5.2.4 Design of Parameters and Optimization

Generally, the design parameters of the mechanism have a great influence on its performance and physical size. The design objectives of the constant-force stage involve a good constant-force property, a constant-force stroke of over 100 μm, and a compact physical structure. In addition, the stage will be developed for use with an inverted microscope. Thus, the size of the stage is constrained by the workspace, that provides one boundary condition for the parameter design. In accordance with the design goal, the design order is planned as the displacement amplifier, negative-stiffness mechanism, positive-stiffness mechanism, and remaining structures, sequentially.

The parametric design for the displacement amplifier relies on the parameters of PZT. In this work, one PZT (model: P-885.91, from Physik Instrumente GmbH) is chosen to actuate the constant-force stage. The maximum stroke of the PZT is 32 μm. To generate a large enough output displacement (over 100 μm) for the stage, the amplification ratio for the bridge-type amplifier is no less than 3.2. Taking into account that the stage material is chosen as ABSplus for 3D printing, the stage structure exhibits a low stiffness. It is easy to be deformed in the working process, that induces the lost motion and reduces the actual magnification ratio. In consideration of the existence of lost motion, the theoretical value of the amplification ratio is designed as sufficiently large (e.g., 7.0 in this work). After determining the amplification ratio, the parameters of the displacement amplifier are obtained as shown in Table 5.1.

Afterward, the parameters of the negative-stiffness mechanism are designed. Generally, the size design is a compromise process. In particular, if the length is too short, it requires a larger actuation force to make it buckle. This needs a motor with large enough driving force, causing unnecessary high cost. On the other hand, if the length is too long, it results in a larger size for the stage. Additionally, as limited by the fabrication capability, a too thin beam cannot be machined. Owing to the size limit, the length of the bistable beam is constrained within 40 mm. In consideration of the

Table 5.1 Main structural parameters of the constant-force stage

Component	Parameter	Value	Unit
Amplifier	l	23	mm
	t	3	mm
	r	0.5	mm
	w	3	mm
Bistable beam	l	32.5	mm
	t	1	mm
	w	3	mm
	θ	4	degree
Leaf flexure	l	19.5	mm
	t	1	mm
	w	3	mm

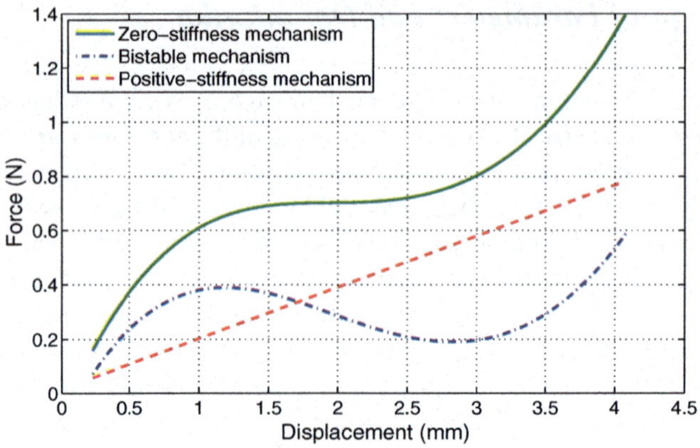

Fig. 5.5 Analytical model results of three types of mechanisms

machining capability, a lower limit of 0.5 mm is assigned to the in-plane width of the bistable beam. In order to guarantee the loading capability, the minimum out-of-plane thickness is set as 2 mm. In addition, to alleviate the influence of machining error, a larger than 2° inclination angle is assigned to the bistable beam. Besides, taking into account the performance of bistable beam, the stroke of negative-stiffness mechanism should be greater than 100 μm. By this, the constant force can attain within 100 μm. To guarantee the magnitude of the constant force near 1.4 N, the force value at the point of buckling should lie in the range from 0.3 to 0.8 N.

The aforementioned design constraints are concluded below.

$$
\begin{aligned}
&0 < l \leq 40\,\text{mm} \\
&t \geq 0.5\,\text{mm} \\
&\omega \geq 2\,\text{mm} \\
&\theta \geq 2^\circ \\
&d \geq 300\,\mu\text{m} \\
&0.3\,\text{N} \leq F \leq 0.8\,\text{N}
\end{aligned}
\tag{5.11}
$$

By taking into account all of the design issues, the optimal parameters are determined with the help of genetic algorithm (GA)-based optimization approach. The optimized parameters are given in Table 5.1. Inserting these parameters into the elliptic integral model, we can obtain the relationship between the force–displacement relationship for the beam, which is depicted in Fig. 5.5 (dash–dot curve). We can observe that the bistable beam exhibits buckling at the displacement of 1.2 mm and the stiffness value becomes negative.

Then, the parameters of the positive-stiffness mechanism are determined. As discussed earlier, the MLF is adopted as the positive-stiffness mechanism. It is integrated with negative-stiffness bistable beams to form a zero-stiffness mechanism. Thus, the elliptic integral model is firstly used to determine the parameters of MLF. The stiffness of the negative-stiffness mechanism is computed from the result as given in Fig. 5.5. Afterward, substituting the result into Eq. (5.10) and synthesizing each kind of design factor (i.e., manufacture capability, compactness of the stage, and so on), the parameters are determined initially. Then, the initial parameters are used to construct the CAD model in simulation study. Based on the simulation results, the parameters are finely adjusted to generate a better constant-force property. In accordance with the foregoing design procedures, the parameters of MLF are determined as shown in Table 5.1.

The force–displacement relationship of the MLF is depicted in Fig. 5.5 (dotted curve). We can observe that the application of MLF eliminates the phenomenon of stress stiffening.

Afterward, the positive-stiffness mechanism and negative-stiffness mechanism are combined together to obtain the zero-stiffness mechanism. Substituting the design parameters into the analytical model, we can yield the results as illustrated in Fig. 5.5 (solid curve). We can see that the magnitude of constant force is around 0.69 N and the constant-force stroke is about 1 mm.

5.2.5 *Design of the Layout*

The entire constant-force stage system is composed of the driving unit, sensor unit, and guiding unit. The driving unit includes the PZT and displacement amplifier. The sensor unit involves a displacement sensor and a force sensor, where the latter is adopted to measure the driving force for the stage. Through the guiding unit, the actuation force provided by the driving unit is transmitted into the output end. In addition to the role of motion transmission, this unit also functions as the zero-stiffness mechanism of the stage. In practice, a preloading is needed to make the stage initially work in the constant-force stroke.

The constant-force stage possesses a symmetric architecture. Figure 5.6a displays the layout design of the stage mechanism. As mentioned above, the selected ABSplus material has a relatively low stiffness. As a result, no matter what the size of the support structure is, certain deformation will be presented on the support structure, which will degrade the performance of the stage. For overcoming this issue, multiple bolts are employed to firmly constrain the freedoms of the support mechanism.

As for the guiding unit, the dominant points of force bearing are the connections between the guiding mechanism and support mechanism. Thus, the bolts #5, #6, #12, and #13 are utilized to constrain the mechanism. In addition, to restrict the translational freedom, bolts #2 and #3 are employed. Bolts #1 and #4 are used to eliminate the rotational freedom. Similarly, bolt #9 is utilized to bear the force and bolts #7, #8, #10, and #11 are adopted to restrict the translational and rotational freedoms. Figure 5.6b depicts a CAD model for the assembled stage.

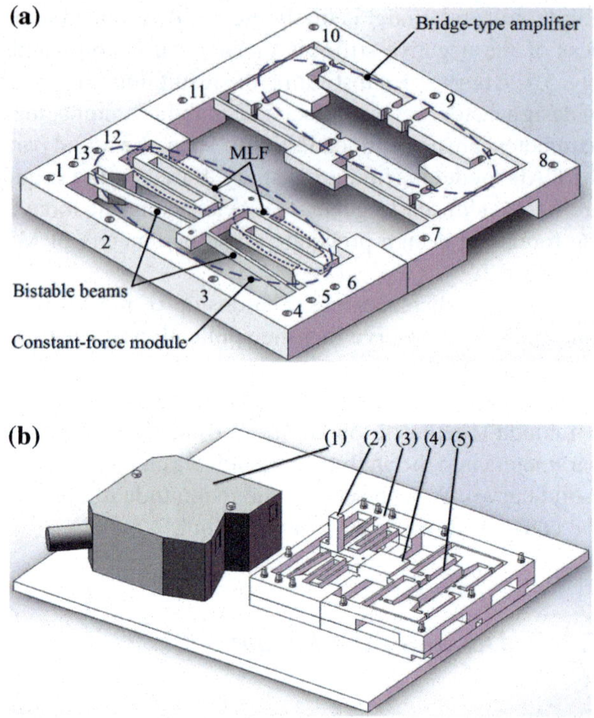

Fig. 5.6 **a** Layout of the stage mechanism with 13 through-holes for fixing; **b** CAD model of the assembled stage with (1) laser displacement sensor, (2) laser sensor target, (3) stage mechanism, (4) force sensor, and (5) PZT actuator

5.3 Performance Evaluation with FEA Simulation

To design a constant-force stage system with desired characteristics, the structural parameters are carefully determined, as shown in Table 5.1. Additionally, for testing the performance of the whole stage, FEA simulation study is carried out in this section.

5.3.1 Amplification Ratio Assessment

The stage's effective stroke refers to its constant-force stroke. The design objective for the stage's stroke is larger than 100 μm . Hence, the amplification ratio is an important factor of the stage, that determines whether the stage can cover the constant-force stroke. In view of Eq. (5.4) and Table 5.1, the amplification ratio is calculated as 7.6. Figure 5.7a shows the FEA simulation result, which gives an amplification ratio of 6.7. The discrepancy between the two results is induced by the assumption made in the analytical model, where the middle parts are supposed as rigid body. Yet, they are not ideally rigid bodies in practice. Thus, the actual amplification ratio is

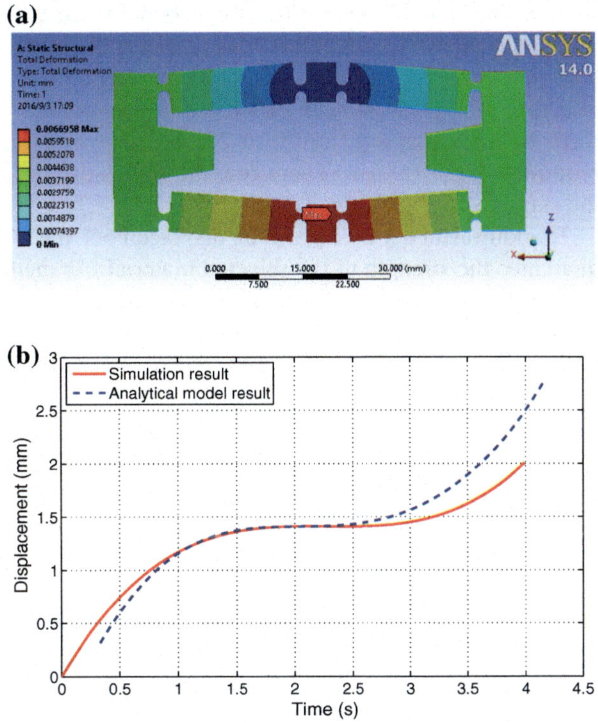

Fig. 5.7 **a** Simulation results of deformation of the displacement amplifier; **b** constant-force property of the zero-stiffness structure

overestimated by the analytical model. On the contrary, all kinds of deformations are considered in the FEA simulation, that is much closer to practical situation. Even so, it is found that the resulted amplification ratio meets the requirement of the stage stroke well.

5.3.2 Actuation Force and Stress Evaluation

The constant force output is an important characteristic of the designed stage system. It is necessary to assess the magnitude of the constant force in order to guarantee that the chosen actuator can drive it. By substituting the design parameters into the FEA model, the force–displacement relationship is obtained as depicted in Fig. 5.7b. We can see that the constant-force stroke of the stage is around 700 μm ranging from 1.7 to 2.4 mm. The actuation force of one zero-stiffness mechanism is about 1.4 N. As compared with the analytical model result, the curves of the two different models display similar tendency, but they do not overlap exactly. Generally, the FEA model result is much closer to real value. Nevertheless, the elliptic integral model is more useful for the design purpose owing to its calculational efficiency. Hence, the elliptic integral model is adopted to determine the preliminary parameters of the bistable beams. Furthermore, the parameters are finely tuned based on FEA model results.

The adopted PZT can offer the 950-N maximum actuation force and 32-μm maximum displacement. Thus, the property of output force for the displacement amplifier can be generated. The constant force of the stage is around 1.4 N. By referring to Fig. 5.8, we can observe that the actuator structure provides an output force of 1.4 N ranging from 0 to 209.6 μm, where range is much larger than the requirement of 100 μm. Therefore, the PZT's driving force is sufficient to actuate the stage.

In addition, under the maximum displacement, the maximum stress is evaluated to guarantee the strength of the selected material. As mentioned earlier, by taking into account the preloading, the maximum displacement about 3 mm is derived. To protect the stage without damage, the desired maximum displacement is increased by 10%. That is, $3 \times (1 + 10\%) = 3.3\,\text{mm}$. As shown in Fig. 5.9, the obtained maximum

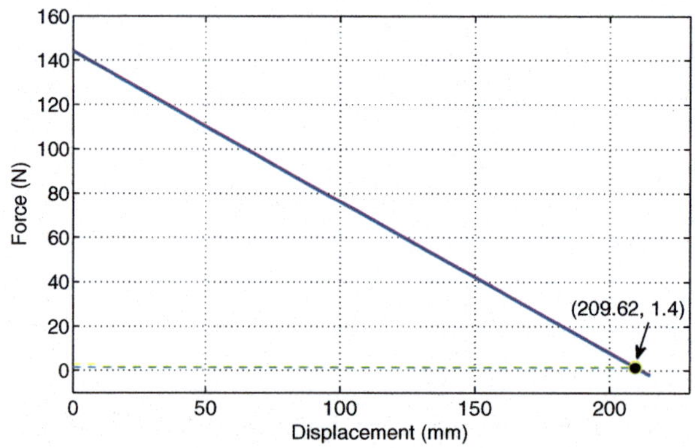

Fig. 5.8 Output force property of the displacement amplifier

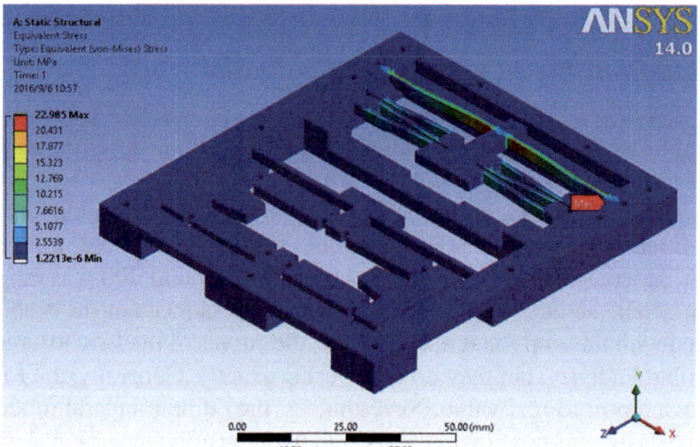

Fig. 5.9 Simulation results of stress distribution for the injector mechanism

displacement by the simulation study is 3.3 mm. The value of the maximum stress is less than the yield strength of the used ABSplus material. Therefore, during the work process of the stage, no plastic damage will be induced.

5.4 Performance Testing by Experimental Study

In this section, a prototype constant-force mechanism is fabricated and its performance is tested by carrying out experimental studies.

5.4.1 Prototype Fabrication

For experimental testing, the designed constant-force stage has been fabricated by a 3D printer with ABSplus material. The prototype is shown in Fig. 5.10. To prevent the beam undergoing too large deformation that may induce plastic deformation, a stroke limiter is adopted. In addition, the stroke limiter functions as reinforcing ribs, that can avoid the deformation for the device during the assembly process.

As depicted in Fig. 5.10, the constant-force stage is actuated by a PZT actuator, that is driven by a high-voltage amplifier (model: EPA-104, from Piezo System, Inc.). The output displacement of stage is measured by using a laser displacement sensor (model: LK-H055, from Keyence Corp.) with the resolution of 50 nm. In addition, a force sensor (model: LSB200, from FUTEK Advanced Sensor Technology, Inc.) is installed between the motor and the driving end of the stage for measuring the

Fig. 5.10 Experimental setup for performance testing of the entire constant-force stage mechanism

driving force of the stage. The controller is developed by employing NI CRIO-9075 hardware (from National Instruments (NI) Inc.). The whole experimental setup is placed on a vibration isolation table to avoid external disturbances.

To verify the effectiveness of the developed constant-force stage, experimental testings are conducted in the following discussion for three purposes, i.e., to test the constant-force performance of the stage, to test the constant-force repeatability, and to test the advantage of the constant-force stage versus conventional one.

5.4.2 Testing Result of Constant-Force Performance

The constant-force capacity is the main characteristic of the developed stage. In order to make the stage initially work in the constant-force stroke, a preloading is applied as an initial displacement. To determine the preloading displacement and the constant-force stroke of the structure, the zero-stiffness mechanism is tested first. The experimental setup is shown in Fig. 5.11. An XYZ stage is chosen as the actuator owing to its large enough output displacement. A translational motion is provided by the XYZ stage to actuate the mechanism to translate in the working direction. The driving force and output displacement are measured by the force sensor and laser displacement sensor, respectively.

The force–deflection relationship of the constant-force mechanism is generated as shown in Fig. 5.12, where the FEA simulation and analytical model results are presented for comparison purpose. The experimental result for the constant-force magnitude is around 1.38 N, that is 1.4% lower than the simulation result (1.4 N). The

Fig. 5.11 Experimental setup for performance testing of the zero-stiffness mechanism

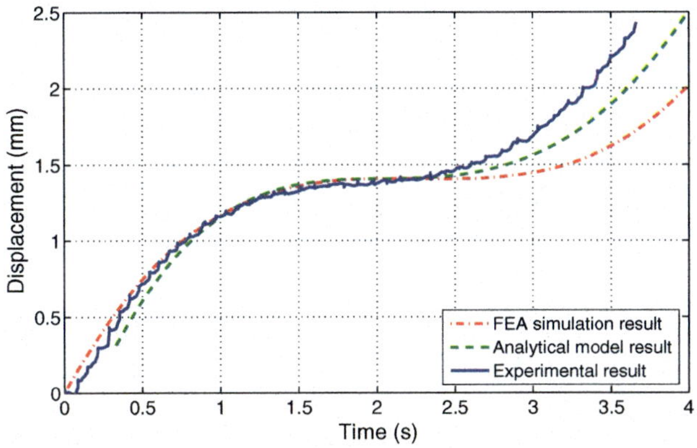

Fig. 5.12 Evaluation results of constant-force property for the injector mechanism

experimental result of the motion stroke is about 800 μm, which ranges from 1.4 to 2.2 mm. The experimental result is 14.2% larger than the simulation result (700 μm) ranging from 1.7 to 2.4 mm. Considering the experimental result, the preloading displacement can be derived as 2 mm. It is observed that the variation tendencies of the experimental and simulation results are similar to each other. The discrepancy between the results is mainly attributed to the fabrication errors as discussed earlier.

Then, the performance of the whole constant-force stage mechanism is examined. The PZT is driven by a sinusoidal signal to generate the maximum output displacement for the stage. Figure 5.13a shows the results. We can observe that the maximum stroke of the stage mechanism is about 138 μm, that is 38% larger than the design objective of 100 μm. Thus, the driving force and output displacement for the stage mechanism satisfy the application requirement.

Moreover, a triangle signal is adopted to actuate the stage, and the force–displacement relationship of the stage is generated as depicted in Fig. 5.13b. We can see that the constant force exhibits certain variation. In particular, as the displacement increases, the magnitude of constant force has a slight rising trend. Specifically, the force magnitude has been increased about 0.04 N with an output displacement of 138 μm. The fluctuation of the force is about 2.9%. Hence, a nearly constant force magnitude is obtained in the displacement range.

5.4.3 Repeatability Testing Result

Repeatability is a crucial performance factor of the positioning stage. The constant-force mechanism is expected to produce a constant force magnitude. Moreover, it is desired to produce the same value of the force during the back-and-forth movement with multiple tests, which indicate the output stability for the constant-force mechanism.

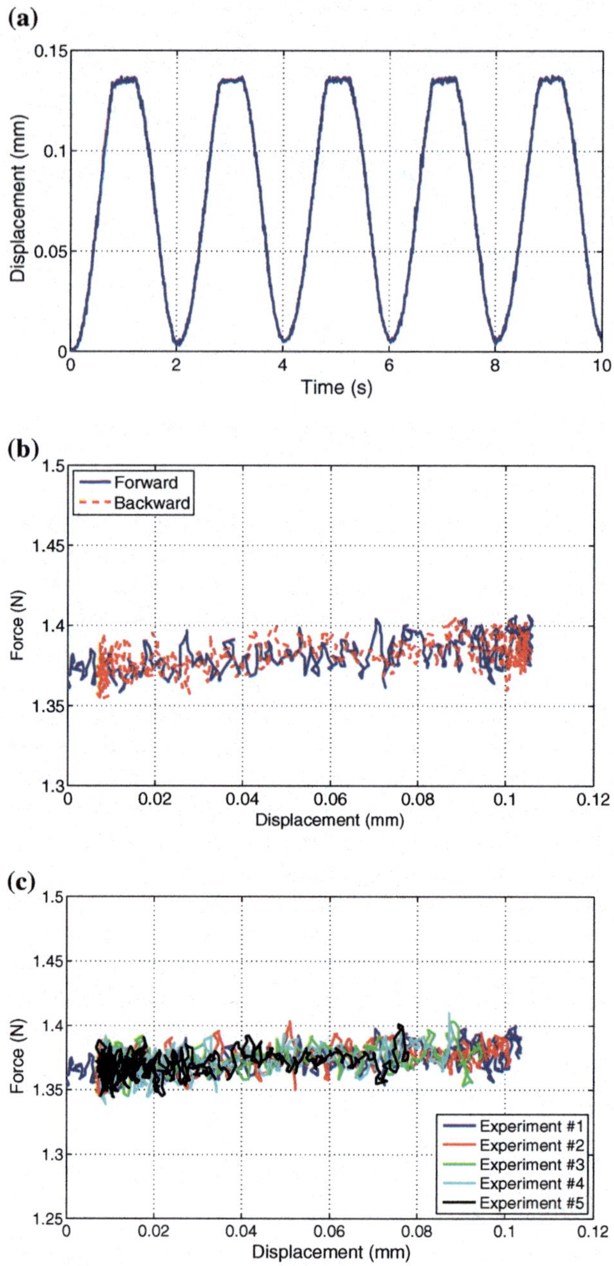

Fig. 5.13 Experimental results of **a** output displacement, **b** back-and-forth motion testing, and **c** multiple testings

We can observe from Fig. 5.13b that the round-trip curves are almost overlapped with each other. The discrepancy between these two curves is smaller than 0.01 N, that indicates a less than 0.7% hysteresis width. In addition, to evaluate the performance of multiple travels of the stage, the former test has been repeated by five times and the results are illustrated in Fig. 5.13c. We can see that the magnitude of the constant force floats around 1.38 N with an error less than ±0.01 N. This reveals a repeatable constant-force output for the developed stage.

5.4.4 Comparison Experimental Result

The developed constant-force stage is able to reduce the actuation force. In this section, a comparison experiment is performed to demonstrate the advantages of the designed stage. To make a fair comparison, a conventional stage is fabricated with the parameters identical to those of the constant-force stage except for the inclined angle of the guiding beams. In particular, the two guiding beams are arranged with an inclined angle, which results in bistable property for the constant-force mechanism, whereas these two guiding beams function as positive-stiffness mechanism in the conventional stage. The prototype is also fabricated using 3D printer, as shown in Fig. 5.14.

To test the performance of the conventional stage, the same sinusoidal signal is applied to PZT to drive the conventional stage to translate in the working direction. The force and displacement data are acquired and depicted in Fig. 5.15. We can observe from Fig. 5.15b that the conventional stage requires an average driving force of about 2.25 N, that is 63% greater than the developed constant-force stage (1.38 N) as given in Fig. 5.13b. As the displacement increases up to 55 μm, the actuation force of the conventional stage is increased about 0.13 N, that is more than three times larger than 0.04 N of the developed constant-force stage.

Therefore, the performed experimental studies have verified the advantages of the developed constant-force stage. The constant-force mechanism is able to reduce the

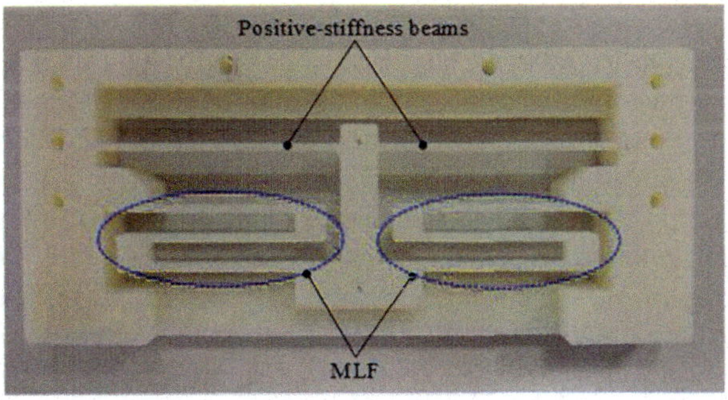

Fig. 5.14 Prototype of the fabricated conventional positive-stiffness stage

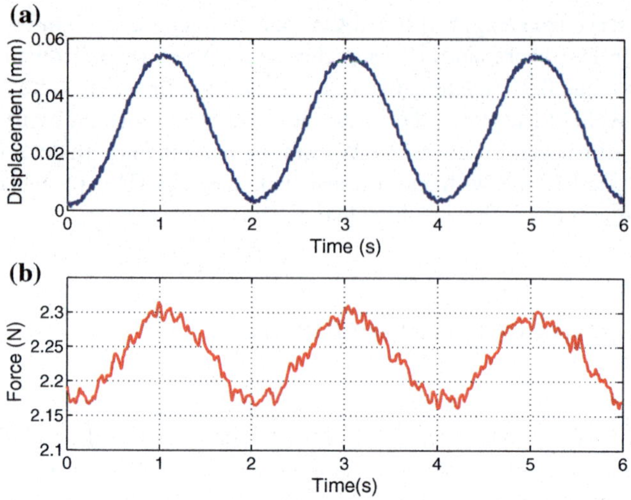

Fig. 5.15 Experimental results of the conventional positive-stiffness stage

driving force (as well as the force fluctuation magnitude) and increase the output displacement indirectly.

5.5 Applications in Biological Micromanipulation

The fabricated constant-force stage mechanism can be adopted in different applications. For illustration, it is applied to biological cell micromanipulation in this section.

5.5.1 Experimental Setup

The fabricated constant-force stage is adopted as a cell microinjector to investigate the mechanical property of the biological cell in this work. In particular, in order to demonstrate the performance of the developed constant-force mechanism, the biological cells are selected as crab eggs (with diameter of 1160 μm) in experimental studies. The experimental setup is shown in Fig. 5.16a. To monitor the cell deformation process, an inverted microscope (model: IX81, from Olympus Corp.) is employed. A pico-injector (model: PLI-100A, from Harvard Apparatus) is adopted to immobilize the cell with the help of negative pressure. Figure 5.16b depicts a close-up view of the experimental setup. The developed constant-force injector is fixed on an XYZ micromanipulator (model: MP-285, from Sutter Instrument Corp.),

(a)

(b)

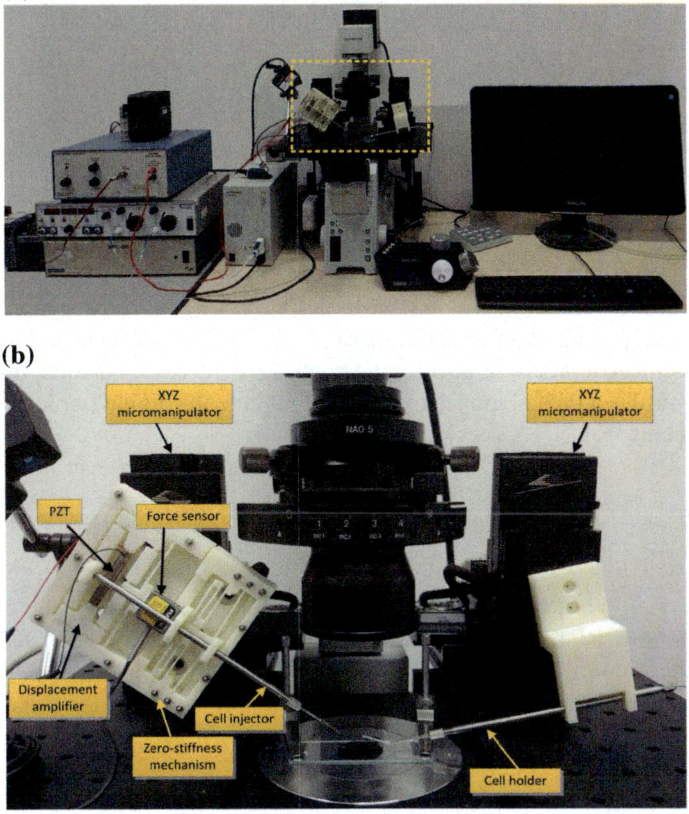

Fig. 5.16 Experimental setup for application study. **a** Overall view of the bio-micromanipulation system; **b** close-up view of the cell injector and holder

and the crab eggs are immobilized through a cell holder via negative pressure. Two types of micropipette are utilized in the experiments. One micropipette is adopted to test the cell's mechanical property. It owns the inner and outer diameters of 0.5 and 1.0 mm, respectively. The other micropipette (outer diameter is 1.0 μm) is used to perform the cell microinjection manipulation (Fig. 5.17).

5.5.2 *Controller Design*

Due to the use of piezoelectric actuator, the injector system exhibits hysteresis nonlinearity. It is notable that the constant-force property of the injector is not influenced by the hysteresis effect. Even so, a closed-loop motion control is required to achieve

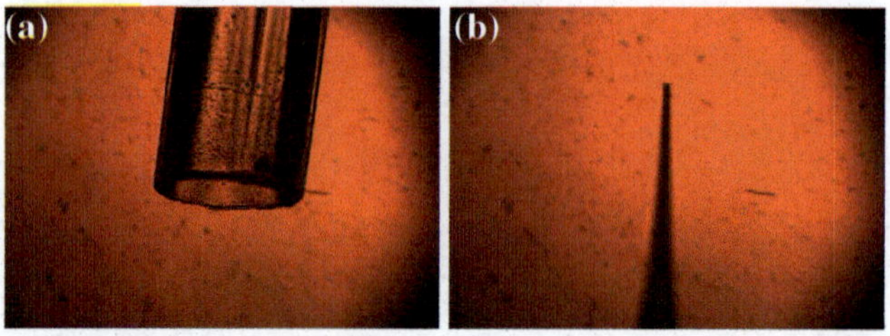

Fig. 5.17 Glass micropipettes for **a** mechanical property testing and **b** cell injection

a precision positioning of the injector. In this work, a PID control algorithm is implemented to achieve a precision positioning control for the constant-force microinjector. PID control is a typical model-free control scheme. Its digital implementation is expressed below.

$$u_n = K_p \cdot e_n + K_i \cdot \sum_{i=0}^{n} e_n + K_d \cdot (e_n - e_{n-1}) \tag{5.12}$$

where $e_n = r_n - y_n$ is the position error with y_n and r_n describing the actual and desired positions in the nth time step, respectively. In addition, K_p, K_i, and K_d represent the proportional, integral, and derivative gains, respectively. A desirable operation of the PID control strategy relies on suitable tuning of the three gains. In this work, the Ziegler–Nichols method is employed owing to its popularity.

5.5.3 Mechanical Property Testing of Biological Cell

In this work, the mechanical property of crab egg is studied with the help of the proposed constant-force microinjector device. The calibration is first conducted, and the initial reading of the force sensor is assigned as zero after applying the preloading. Then, a sinusoidal signal is applied to actuate the constant-force stage to translate in the working direction. Figure 5.18 shows the experimental results. The snapshots of cell deformation procedure are depicted in Fig. 5.19.

We can observe that the injector does not contact the cell before the displacement of 0.04 mm. The contact is established between the injector tip and the cell at the displacement of 0.04 mm. The cell experiences a deformation in the time period of 0.4–0.9 s, which is imposed by the microinjector tip. As the constant-force injector requires a nearly constant driving force, the force sensor readings do not change until the microinjector tip contacts the cell wall. Thus, the variation on the output force is entirely attributed to the cell's deformation. Hence, the use of the developed constant-force stage facilitates the direct study of the cell's mechanical property without employing a complex signal processing method.

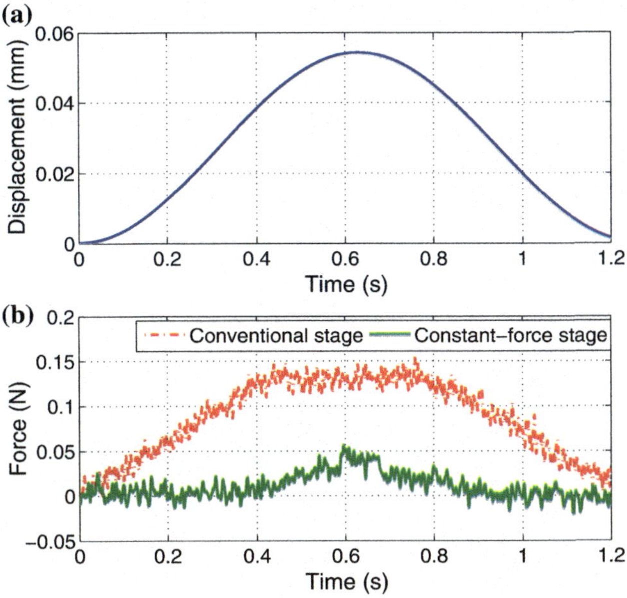

Fig. 5.18 Mechanical property testing results with small deformation of cell

Fig. 5.19 Snapshots of cell manipulation in mechanical property testing

For comparison, the constant-force mechanism is replaced by a conventional flexure-based mechanism, which functions as a positive-stiffness guiding mechanism. The foregoing experimental study is repeated, and Fig. 5.18b (dash–dot curve) depicts the experimental result. By examining the two curves in Fig. 5.18b, we can observe that the output force of the conventional mechanism cannot reflect the force that is exerted to the cell. The reason is that the conventional flexure-based mechanism does not exhibit a constant driving force. The force fluctuation is larger than that of the developed constant-force mechanism. As a result, the driving force and the external force exerted to the cell are mixed together. Hence, the force sensor readings are complex signal. To examine the mechanical property of the cell, the conventional mechanism demands a signal processing technique, that complicates the analysis process.

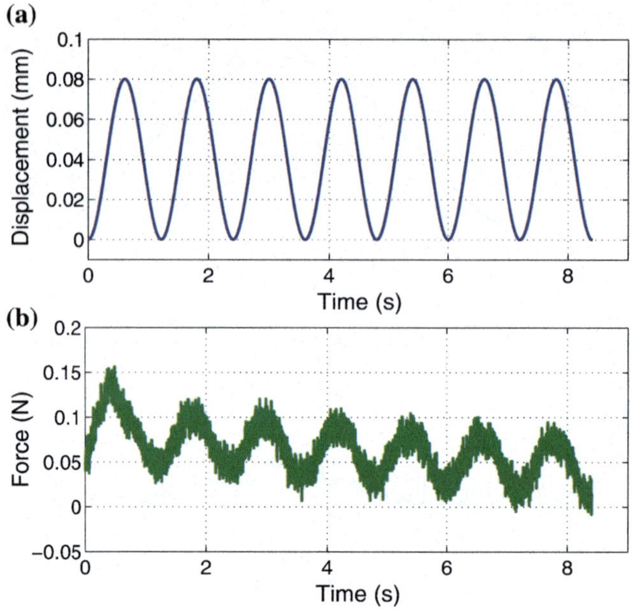

Fig. 5.20 Mechanical property testing results with large deformation of cell

Additionally, the cell's mechanical property can be tested by using the same glass micropipette when a large deformation is experienced. In the experiments, an XYZ micromanipulator is actuated to translate the constant-force injector adjacent to the target position, where the injector tip just touches the cell wall. And then, the PZT-actuated microinjector is driven to follow a 80-μm sinusoidal trajectory, as illustrated in Fig. 5.20a. By this way, a large deformation of 80-μm is exerted on the cell, repetitively. Figure 5.20b shows the force signal. We can observe that force exerted by the cell exhibits a descending trend. This phenomenon implies that the cell gradually suffers from a plastic deformation when it experiences a large deformation.

5.5.4 Experimental Testing of Cell Injection

One of the dedicated applications of the developed constant-force mechanism is cell microinjection. Considering that the stroke of the stage is about 138 μm, the developed injector can be used to inject small cells (less than 500 μm). Given larger cells (e.g., crab eggs), the injector's stroke does not satisfy the requirement of standard microinjection. Under this scenario, the fabricated mechanism can act as a force sensor in the cell microinjection operation, that is performed with a larger-stroke actuation. As compared with conventional force sensor, the new force sensor possesses some advantages due to the zero-stiffness property, which is demonstrated in the following.

In the experimental study, the fabricated constant-force microinjector is mounted onto an XYZ micromanipulator. The microinjector is driven by a constant voltage signal, and the XYZ micromanipulator is actuated by a ramp signal for cell microinjection. Figure 5.21 (solid curve) depicts the experimental result. In addition, Fig. 5.22 shows the snapshots of the cell deformation procedure. We can see that the injector tip does not contact the cell before the displacement of 1.35 mm. After this displacement, the injector tip begins to pierce the cell wall. And then, the penetration is continued during the displacement range of 1.35–1.85 mm. After the displacement of 1.85 mm, the cell is punctured by the injector tip. Hence, the stroke of injection is 0.50 mm.

Generally, the introduced new force sensor exhibits two merits. First, the force influence of the guiding mechanism is prevented, and force experienced by the cell is directly reflected by the sensor output signal as discussed earlier. Second, the sensor readings truly describe the relationship between the force and displacement of the cell in the microinjection process, which is enabled by mounting the sensor on the injector's side. To demonstrate this effect, an additional experiment is carried out to conduct a comparison study. Specifically, Fig. 5.21 (dashed curve) shows the experimental result when the force sensor is fixed on the holder's side. By examining the

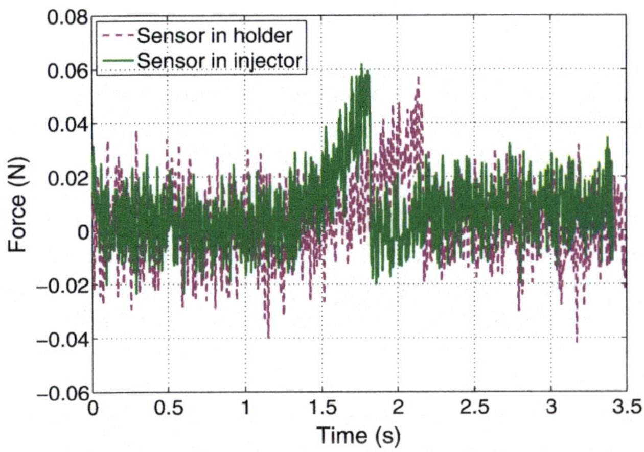

Fig. 5.21 Cell injection results with different configurations of the force sensor

Fig. 5.22 Snapshots of cell microinjection manipulation

two curves in Fig. 5.21, we can observe that the strokes of injection are not identical. In particular, the second experiment requires a larger stroke, that is performed with the sensor mounted on the holder's side.

In order to remove the influence caused by cell variation, the two foregoing experiments adopt the cells with almost identical size. The cells are selected carefully and placed in the same environment of culture media. Moreover, the experimental studies have been conducted by multiple times, which provide the same conclusion. By getting rid of other influential issues, we can deduce that the discrepancy between the results of these two experiments is mainly caused by the mounting positions of the force sensor. The force sensor's operation principle is based on the relationship between the force and deformation. That is, the sensor exhibits an elastic deformation when a force is exerted on it. Thus, when the force sensor is fixed on the injector's side, the sensor's deformation can be easily measured by the adopted displacement sensor (i.e., laser displacement sensor) in this work. Nevertheless, if the force sensor is mounted on the cell holder's side, the sensor's deformation cannot be detected easily. In reality, this sensor deformation is commonly neglected, that degrades the accuracy of the experimental result. Hence, the superiority of the proposed force sensor is evident in comparison with conventional one.

5.6 Conclusion

This chapter presents the design, analysis, and experimental study of a flexure-based constant-force mechanism. Analytical model of the zero-stiffness mechanism has been derived and validated by carrying out simulation study with FEA. Simulation results exhibit a reasonable actuation force and maximum stress of the mechanism. Experimental results show that the developed prototype offers a constant output force of 1.38 N in the output displacement of 138 μm. A slight fluctuation of 2.9% of the driving force is observed. Additionally, the mechanism exhibits a negligible hysteresis, which is lower than 0.7% of the maximum force in back-and-forth testing. Multiple tests reveal a fluctuation of less than ±0.01 N for the constant force, which indicates a fine repeatability. Further comparison study has been conducted with both conventional mechanism and developed constant-force mechanism. Results confirmed the advantages of constant-force mechanism in terms of reducing the driving force and its fluctuation and enlarging the output displacement. To demonstrate the effectiveness of the developed constant-force mechanism, it is adopted as a cell injector in biological cell micromanipulation, which avoids the force influence of the guiding mechanism and enables a more accurate force sensing in the manipulation.

References

1. Chen, G., Gou, Y., Zhang, A.: Synthesis of compliant multistable mechanisms through use of a single bistable mechanism. J. Mech. Des. **133**(8), 081007 (2011)

2. Chen, Y.H., Lan, C.C.: An adjustable constant-force mechanism for adaptive end-effector operations. J. Mech. Des. **134**(3), 031005 (2012)
3. Choi, K.B., Kim, D.H.: Monolithic parallel linear compliant mechanism for two axes ultraprecision linear motion. Rev. Sci. Instrum. **77**(6), 065106 (2006)
4. Choi, Y.J., Sreenivasan, S., Choi, B.J.: Kinematic design of large displacement precision xy positioning stage by using cross strip flexure joints and over-constrained mechanism. Mech. Mach. Theory **43**(6), 724–737 (2008)
5. Goncalves, F.D., Finnegan, P.F., Sigman, G., Brown, M.V.: Caster wheel with constant force mechanism (2015). US Patent 20,150,274,495
6. Gu, G.Y., Zhu, L.M., Su, C.Y., Ding, H., Fatikow, S.: Modeling and control of piezo-actuated nanopositioning stages: a survey. IEEE Trans. Autom. Sci. Eng. **13**(1), 313–332 (2016)
7. Holst, G.L., Teichert, G.H., Jensen, B.D.: Modeling and experiments of buckling modes and deflection of fixed-guided beams in compliant mechanisms. J. Mech. Des. **133**(5), 051 (2011)
8. Howell, L.L.: Compliant Mechanisms. Wiley, New York (2001)
9. Klein, R.J.: Constant force compression tool (2013). US Patent App. 13/746,929
10. Lamers, A.J., Sanchez, J.A.G., Herder, J.L.: Design of a statically balanced fully compliant grasper. Mech. Mach. Theory **92**, 230–239 (2015)
11. Lan, C.C., Cheng, Y.J.: Distributed shape optimization of compliant mechanisms using intrinsic functions. J. Mech. Des. **130**(7), 072304 (2008)
12. Liu, Y., Shan, J., Gabbert, U.: Feedback/feedforward control of hysteresis-compensated piezoelectric actuators for high-speed scanning applications. Smart Mater. Struct. **24**(1), 015012 (2015)
13. Liu, Y., Zhang, Y., Xu, Q.: Design and control of a novel compliant constant-force gripper based on buckled fixed-guided beams. IEEE/ASME Trans. Mechatron. **22**(1), 476–486 (2017)
14. Malloy, A.L., Radzik, J., Dean, M., Hauver, B.C., Knaus, G.A., Thomas, C.: Constant force coaxial cable connector (2012). US Patent RE43,832
15. Polit, S., Dong, J.: Design of high-bandwidth high-precision flexure-based nanopositioning modules. J. Manuf. Syst. **28**(2), 71–77 (2009)
16. Putra, A.S., Huang, S., Tan, K.K., Panda, S.K., Lee, T.H.: Design, modeling, and control of piezoelectric actuators for intracytoplasmic sperm injection. IEEE Trans. Control Syst. Technol. **15**(5), 879–890 (2007)
17. Qiu, J., Lang, J.H., Slocum, A.H.: A curved-beam bistable mechanism. J. Microelectromech. Syst. **13**(2), 137–146 (2004)
18. Todd, B., Jensen, B.D., Schultz, S.M., Hawkins, A.R.: Design and testing of a thin-flexure bistable mechanism suitable for stamping from metal sheets. J. Mech. Des. **132**(7), 071011 (2010)
19. Wang, J.Y., Lan, C.C.: A constant-force compliant gripper for handling objects of various sizes. J. Mech. Des. **136**(7), 071008 (2014)
20. Wang, P., Xu, Q.: Design and testing of a flexure-based constant-force stage for biological cell micromanipulation. IEEE Trans. Autom. Sci. Eng. (2017). https://doi.org/10.1109/TASE.2017.2733553
21. Wang, P., Xu, Q.: Design of a flexure-based constant-force XY precision positioning stage. Mech. Mach. Theory **108**, 1–13 (2017)
22. Xie, Y., Sun, D., Tse, H.Y.G., Liu, C., Cheng, S.H.: Force sensing and manipulation strategy in robot-assisted microinjection on zebrafish embryos. IEEE/ASME Trans. Mechatron. **16**(6), 1002–1010 (2011)
23. Xu, Q.: Adaptive discrete-time sliding mode impedance control of a piezoelectric microgripper. IEEE Trans. Rob. **29**(3), 663–673 (2013)
24. Xu, Q., Li, Y.: Analytical modeling, optimization and testing of a compound bridge-type compliant displacement amplifier. Mech. Mach. Theory **46**(2), 183–200 (2011)
25. Zhang, W., Sobolevski, A., Li, B., Rao, Y., Liu, X.: An automated force-controlled robotic micromanipulation system for mechanotransduction studies of drosophila larvae. IEEE Trans. Autom. Sci. Eng. **13**(2), 789–797 (2016)

Chapter 6
Design, Modeling, and Control of a Constant-Force Microgripper

Abstract This chapter proposes the design, modeling, and control of a compliant gripper with a passive-type constant-force mechanism. The constant force output is enabled by combining a positive-stiffness mechanism and a negative-stiffness mechanism. The negative stiffness is produced by a bistable fixed-guided beam with buckling behavior. The developed constant-force gripper allows the generation of constant force via its mechanical structure, which allows the elimination of force control. Analytical modeling and nonlinear finite element analysis (FEA) simulation study are carried out to evaluate the gripper performance. A prototype is developed for experimental study. To achieve a precise positioning of the gripper jaw, a discrete-time sliding mode control strategy is developed on the basis of a nonswitching-type reaching law. The effectiveness of the gripper system is validated by performing experimental studies on grasp-hold-release manipulation of micro-object.

6.1 Introduction

Generally, a robotic gripper can realize the automated grasp-hold-release manipulation, which is widely demanded in the applications including material characterization, biological sample pick-and-place, and micro-object assembly [6, 7]. This chapter introduces the design and development of a gripper device dedicated to micromanipulation and microassembly tasks in laboratory or industry production line. Due to different actuating principles, various grippers have been reported in the literature [9, 18]. In particular, piezoelectric stack actuator (PSA) has been extensively employed thanks to the remarkable advantages in terms of large blocking force, ultrahigh motion resolution, and rapid response speed [14, 22]. Therefore, PSA is adopted for driving the gripper in this work.

For a precise and safe manipulation, the regulation of the grasping force poses a challenge on the gripper design and development. When the gripper jaws contact the target object, the gripping force should be regulated under the target's maximum tolerance without damage. In the literature, lots of grippers have been devised with

Q. Xu, *Micromachines for Biological Micromanipulation*,
https://doi.org/10.1007/978-3-319-74621-0_6

force sensors and controllers through different force control algorithms, e.g., switching force/position controller and impedance control [19]. However, such solutions require both position and force control, that dramatically increase the complexity of the gripper system and degrade its practicability. To overcome this issue, an alternative approach is to use a constant-force mechanism (CFM). By this way, the grasping force can be easily regulated to the desired value without adopting a force controller. The output force can maintain a nearly constant value in the grasp range of motion.

Different approaches can be adopted to generate a constant-force mechanism [8, 15]. Intuitively, a positive-stiffness mechanism with a negative-stiffness mechanism can be combined together to form a zero-stiffness mechanism, which provides a constant-force range. The two sub-mechanisms should be designed to exhibit the stiffnesses with the same magnitude and inverse direction. Based on Hooke's law, the positive stiffness can be easily generated by lots of methods, such as normal thin beam. In addition, the negative stiffness can be offered via different approaches, such as vibrational damping and buckling theory. Nevertheless, there are some dilemmas when the vibrational damping approach is adopted to devise a compliant structure [12]. An alternative method is to make use of the negative-stiffness effect of the bistable mechanism with buckling behavior. In the literature, a constant-force compliant gripper has been presented to handle various sized objects [16]. The mechanism design is composed of a constant-torque mechanism and a lever mechanism. Nevertheless, the beam of irregular curve shape is not easily fabricated precisely. In this chapter, an inclined flexure beam-based bistable mechanism is employed to act as a negative-stiffness mechanism, and a new gripper structure is designed to provide a constant output force.

Even though the advantage of the constant-force mechanism lies in the elimination of force control, motion control is still needed to implement an accurate positioning of the gripper in the process of gripping operation. Especially, a precision motion control is demanded for a piezo-driven device in order to suppress the piezoelectric nonlinearities including hysteresis and creep effects. Typical methods of hysteresis compensation include the use of Preisach model [2, 10], Bouc–Wen model [11, 17], Prandtl–Ishlinskii model [1, 20]. Afterward, an inverse hysteresis model is developed as a feedforward control to cancel the hysteresis effect [3]. Nevertheless, both the amplitude and the frequency of the input voltage determine the final effect of the hysteresis. An accurate modeling of the complicated rate-dependence hysteresis behavior is a time-consuming work.

In this work, the piezoelectric nonlinearity is regarded as a lumped disturbance to the nominal system, and a robust control scheme is developed based on sliding mode algorithm to implement the precise position control of the gripper. Several closed-loop position control experiments are conducted to confirm the effectiveness of the developed gripper system.

6.2 Mechanism Design

Figure 6.1 illustrates the designed gripper with a constant-force mechanism of passive type. The gripper comprises two jaws. While the lower jaw is actuated, the upper one is fixed on the base. The actuated gripper jaw mechanism consists of a constant-force module and an actuation module. The constant-force module is composed of two inclined negative-stiffness beams and two straight positive-stiffness beams. The actuation module contains a bridge type of displacement amplifier, which is used to magnify the stroke of piezoelectric stack actuator (PSA). As a passive of type constant-force module, the function of constant-force output for the gripper is triggered on by the reaction force imposed by the grasped target.

6.2.1 Design of the System Stiffness

In this work, the actuated gripper jaw mechanism includes two modules, i.e., the constant-force module and actuation module. The constant force indicates a zero stiffness for the mechanism. In order to achieve a zero stiffness, the system stiffness is designed in the following.

Firstly, the constant-force module is constructed as a zero-stiffness mechanism. It consists of a positive-stiffness part (a) and a negative-stiffness part (b). These two parts are connected in parallel. Concerning a device with parallel springs, we have

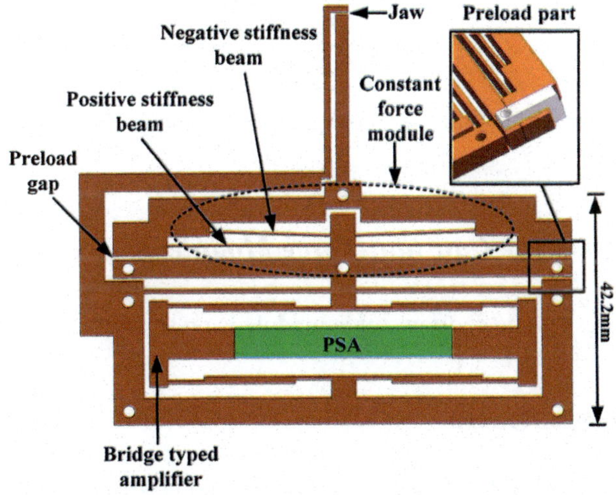

Fig. 6.1 Designed gripper with a passive-type constant-force mechanism

$$F_a = K_a q \quad (6.1)$$

$$F_b = K_b q \quad (6.2)$$

where F_a and F_b denote the force, K_a and K_b represent the stiffness, and q is the displacement of the two parts a and b.

In consideration of the parallel connection, we can derive the relations:

$$F_{module} = F_a + F_b \quad (6.3)$$

$$K_{module} = K_a + K_b. \quad (6.4)$$

Based on Eq. (6.4), the stiffness of the module will become zero if stiffnesses K_a and K_b have the same magnitude and inverse directions. Thus, a positive-stiffness part and a negative-stiffness part are devised to generate a zero-stiffness mechanism.

Secondly, in order to achieve a constant force output for the device, a zero stiffness should be produced by integrating the constant-force module and actuation module. In this work, the actuation and the constant-force modules are connected in series.

Regarding a device with series springs, we have

$$\Delta x = \Delta x_1 + \Delta x_2 \quad (6.5)$$

$$\begin{aligned} F_{gripper} &= K_{gripper} \Delta x \\ &= K_{actuator} \Delta x_1 = K_{module} \Delta x_2 \end{aligned} \quad (6.6)$$

where Δx_1 and Δx_2 are the displacements, and K_{module} and $K_{actuator}$ are the stiffnesses of the constant-force and actuation modules, respectively. Additionally, $K_{gripper}$ and Δx represent the stiffness and displacement of the gripper, respectively.

In consideration of the serial connection, the following stiffness equation is generated.

$$K_{gripper} = \frac{K_{actuator} K_{module}}{K_{actuator} + K_{module}} \quad (6.7)$$

It indicates that the stiffness of the whole gripper mechanism will be zero if K_{module} approaches to zero.

6.2.2 Design of the Constant-Force Module

The design process of the constant-force module is presented in this section.

6.2.2.1 Mechanism Design

According to Eq. (6.4), the constant-force module is composed of a negative-stiffness part and a positive-stiffness part. In this work, the positive stiffness is offered by using straight fixed-guided beams.

The negative-stiffness part is provided by making use of the buckling behavior of the inclined bistable beams. The buckled system consists of different forms. The initially inclined fixed-guided beam is a simple scheme. Additionally, in compliant mechanism design, the beam element exhibits a number of advantages including relatively large deflection and distributed compliance. Thus, it is adopted to offer the negative stiffness in this work.

Concerning the negative-stiffness part, the force (and energy) increases as the displacement increases initially. As the force is increased up to the critical point, the beam is buckled. After that, the stored energy starts to release. As the deformation further increases, the force exerted on the beam decreases. At this moment, the stiffness begins to exhibit opposite direction. In other words, it functions as a negative-stiffness mechanism.

6.2.2.2 Modeling of Bistable Beam

The analytical modeling of the inclined fixed-guided beam has been conducted in different ways in the literature [4, 21]. With reference to the parameters as shown in Fig. 6.2, the model is established based on the Bernoulli–Euler equation below.

$$EI\frac{d\beta}{ds} = -Fx_A \sin\theta + Fy_A \cos\theta + M \tag{6.8}$$

in which the coordinates for the point A are defined as follows.

$$\frac{dx_A}{ds} = \sin\beta \tag{6.9a}$$

$$\frac{dy_A}{ds} = \cos\beta \tag{6.9b}$$

To yield the solution, firstly, A_B and B_B are defined as the horizontal and vertical displacements of the beam experiencing bending deflection. A_B/L and B_B/L

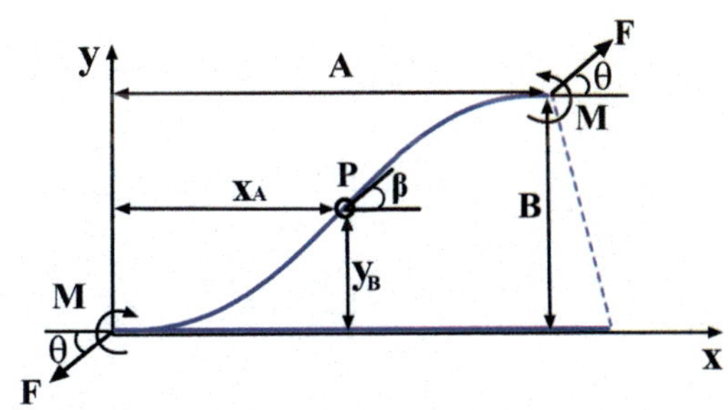

Fig. 6.2 Model parameters of the inclined fixed-guided beam

represent the nondimensionalized horizontal and vertical displacements, respectively. These displacement components are written as follows [4].

$$A_B = \int_0^L (\cos\beta)(EI)^{\frac{1}{2}}[-2F\cos(\beta-\theta)+2C]^{-\frac{1}{2}}d\beta \quad (6.10)$$

$$B_B = \int_0^L (\sin\beta)(EI)^{\frac{1}{2}}[-2F\cos(\beta-\theta)+2C]^{-\frac{1}{2}}d\beta \quad (6.11)$$

$$\begin{aligned}\frac{A_B}{L} &= -[FL^2(EI)^{-1}]^{-\frac{1}{2}} * \{\cos\theta[2E(k,\phi_2)\\ &\quad -2E(k,\phi_1) - 2F(k,\phi_2) - 2F(k,\phi_1)]\\ &\quad +2k\cos\theta(\cos\phi_1 - \cos\phi_2)\} \end{aligned} \quad (6.12)$$

$$\begin{aligned}\frac{B_B}{L} &= -[FL^2(EI)^{-1}]^{-\frac{1}{2}} * \{\sin\theta[2E(k,\phi_2)\\ &\quad -2E(k,\phi_1) - 2F(k,\phi_2) - 2F(k,\phi_1)]\\ &\quad +2k\cos\theta(\cos\phi_1 - \cos\phi_2)\} \end{aligned} \quad (6.13)$$

The first-kind and second-kind incomplete elliptic integrals in Eqs. (6.12) and (6.13) are expressed below.

$$F(k,\phi) = \int_0^{\phi} \frac{1}{(1-k^2\sin\delta)^{\frac{1}{2}}}d\delta \quad (6.14)$$

$$E(k,\phi) = \int_0^{\phi} (1-k^2\sin\delta)^{\frac{1}{2}}d\delta \quad (6.15)$$

with δ being an item in the elliptic integral.

In Eqs. (6.10) and (6.11), the constant parameter C is generated by taking the integral operation of differential equation (6.8), i.e.,

$$C = F(2k^2 - 1) \quad (6.16)$$

In addition, the parameters ϕ and k are defined below.

$$k\sin\phi = \cos\left(\frac{\theta-\beta}{2}\right) \quad (6.17)$$

In Eqs. (6.12) and (6.13), the parameters ϕ_1 and ϕ_2 are related by:

$$\left(\frac{FL^2}{EI}\right)^{\frac{1}{2}} = F(K,\phi_2) - F(K,\phi_1) \quad (6.18)$$

And then, owing to axial deflection (axial strain), the beam's horizontal and vertical displacements are generated as A_A and B_A, respectively. The corresponding nondimensionalized horizontal and vertical displacements are assigned as A_A/L and B_A/L, respectively, that are presented below.

$$A_A = \int_0^L \left[\frac{F\cos(\theta - \beta)}{EA_{area}} \cos\beta \right] ds \tag{6.19}$$

$$B_A = \int_0^L \left[\frac{F\cos(\theta - \beta)}{EA_{area}} \sin\beta \right] ds \tag{6.20}$$

$$\frac{A_A}{L} = \int_0^1 \left[\frac{F\cos(\theta - \beta)}{EA_{area}} \cos\beta \right] ds' \tag{6.21}$$

$$\frac{B_A}{L} = \int_0^1 \left[\frac{F\cos(\theta - \beta)}{EA_{area}} \sin\beta \right] ds' \tag{6.22}$$

At last, the overall displacements in horizontal and vertical directions are computed by:

$$A = A_A + A_B \tag{6.23}$$

$$B = B_A + B_B \tag{6.24}$$

The analytical model of force–displacement relationship is obtained for the buckled beam. Figure 6.3a shows the model result which is obtained with the help of MATLAB software. In this case design, the initial angle of the beam is 2.2°, the length of the inclined fixed-guided beam is 20 mm, the in-plane width is 0.1 mm, and the out-of-plane thickness is 4 mm. These parameters are selected based on the specifications of the employed PSA actuator. During the buckling process, the shapes of the deflected beam are shown in Fig. 6.3b, which illustrates the deflection of the beam clearly.

6.3 Simulation Study with FEA

The performance of the integrated design of the positive-stiffness and negative-stiffness mechanisms is verified by performing simulation study with nonlinear FEA in ANSYS APDL program. Such software can be applied to generate the force–displacement relationship of the integrated structure. Thus, it is adopted to determine the size for the positive-stiffness straight beam in this work.

The FEA model of the constant-force module is illustrated in Fig. 6.4. All the degree-of-freedoms of the four endpoints of the beams are constrained. A vertical force is exerted on the middle point. In the simulation, the mechanism's material is assigned as Al-6061. The constant-force performance can be adjusted by tuning different parameters of the straight beam including the length, in-plane width, and

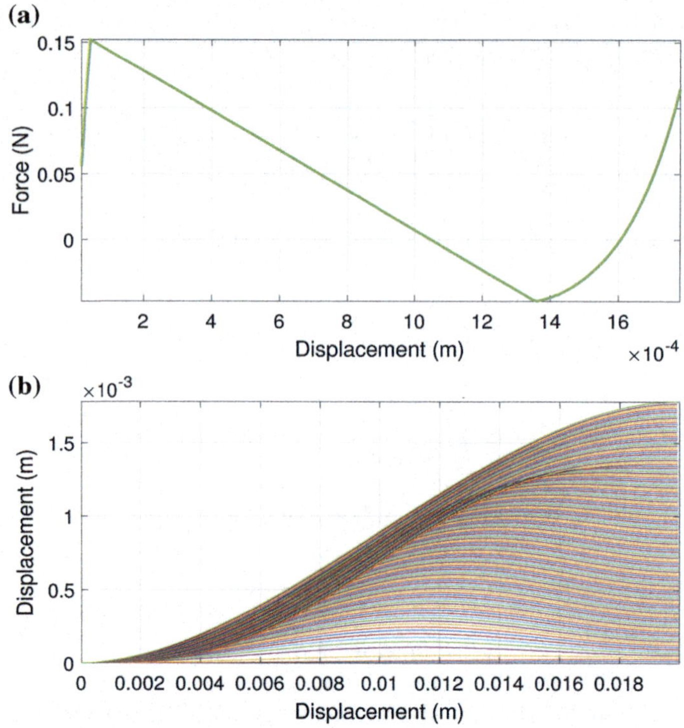

Fig. 6.3 Analytical model results of **a** force–displacement relationship for the negative-stiffness beam and **b** deflected beam shapes

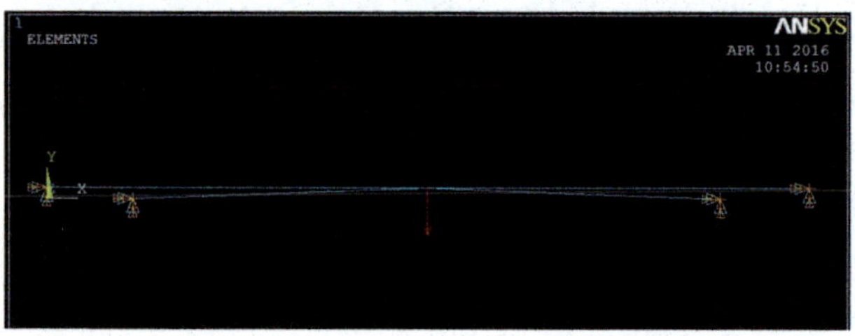

Fig. 6.4 FEA model of the constant-force module

out-of-plane thickness. Riks method is used to solve this buckled problem. Figure 6.5 depicts the simulation results. We can see that the designed module provides a near constant force output of 620 mN in the range of 220 μm. In this case design, the dimensions of the straight beams are chosen as the length of 26 mm, width of 0.1 mm, and thickness of 4 mm. The parameters of the constant-force module are shown in Table. 6.1.

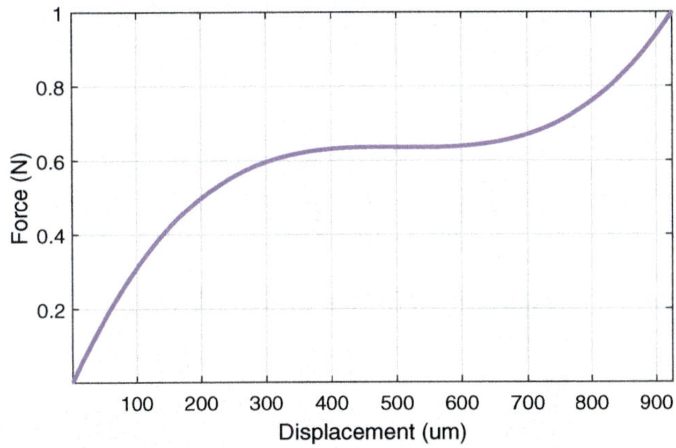

Fig. 6.5 Simulation result of force–displacement relationship for the constant-force module

Table 6.1 Parameters of the constant-force module

Parameter	Value
Length of positive-stiffness beam	26 mm
Width of positive-stiffness beam	0.1 mm
Thickness of positive-stiffness beam	4 mm
Length of negative-stiffness beam	20 mm
Width of negative-stiffness beam	0.1 mm
Thickness of negative-stiffness beam	4 mm
Inclined angle of negative-stiffness beam	2.1°

Additionally, the performance for the actuation module is simulated and the result is displayed in Fig. 6.6. We can see that the actuation module offers an output displacement of 246 μm. In this simulation, the input force is determined by taking into account the output force of the used PSA actuator as well as the stiffness for the driven mechanism.

Referring to Fig. 6.7, the equivalent input force is derived as 475 N by determining the intersection point in this case design. This intersection point is solved by taking into account the two stiffness curves for the actuation module and PSA actuator, where the actuation module's stiffness is generated by the FEA simulation.

In consideration of the output displacement for the actuation module, the initial motion range of the constant-force module (before the constant-force range) is applied to exert a preloading on the constant-force module. In particular, this preloading displacement is applied by tuning the preload gap, which is illustrated in Fig. 6.1. We can see that an L-shape adjuster is used to adjust the preloading displacement. By adopting the L-shape adjuster with different lengths, the preload gap can be adjusted accordingly.

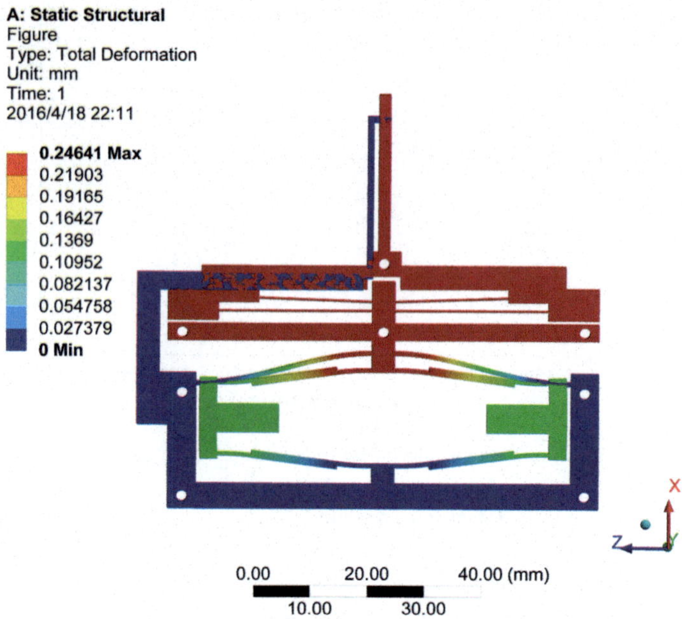

Fig. 6.6 Simulation result of deformation for the actuation module

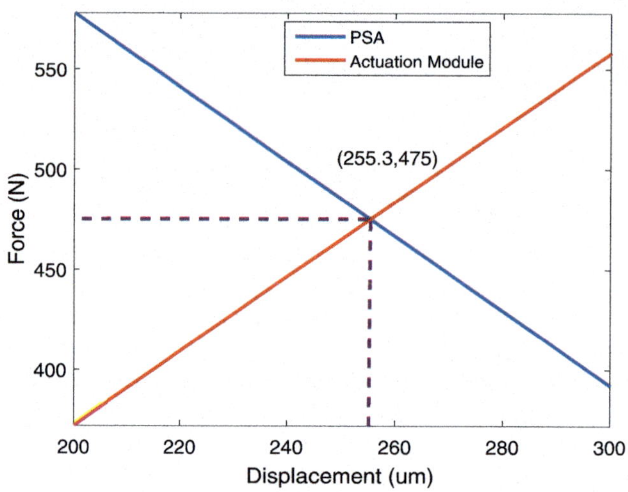

Fig. 6.7 Illustration of actual driving displacement determination

6.4 Design of Sliding Mode Control

In this section, a discrete-time sliding mode motion controller is designed with a nonswitching-type variable structure scheme.

6.4.1 *Nonswitching-Type Reaching Law Design*

In this work, the microgripper is developed by using a flexure-based compliant mechanism which is actuated by a PSA actuator. In order to implement a precision positioning control, the dynamics model is required. The overall dynamics model for the piezo-driven system accompanied with unknown disturbance and hysteresis is derived below.

$$m\ddot{q}(t) + b\dot{q}(t) + kq(t) = du(t) + h(t) + p(t) \tag{6.25}$$

where t represents the time variable, q, m, b, and k are the output displacement, mass, damping coefficient, and stiffness, respectively, u is the input voltage, and d denotes the piezoelectric coefficient. Additionally, h describes the hysteresis effect and p is the lumped perturbation of the system including unmodeled dynamics, model parameter uncertainties, and other unknown items.

The time domain model (6.25) can be converted into the state-space representation as follows.

$$\dot{\mathbf{q}} = \mathbf{A}\mathbf{q} + \mathbf{B}u + \mathbf{P} \tag{6.26}$$

$$y = \mathbf{C}\mathbf{q} \tag{6.27}$$

where the scalar variable y represents the output displacement, and

$$\mathbf{A} = \begin{bmatrix} 0 & 1 \\ -\frac{k}{m} & -\frac{b}{m} \end{bmatrix} \tag{6.28}$$

$$\mathbf{B} = \left[0,\ \tfrac{kd}{m}\right]^T \tag{6.29}$$

$$\mathbf{P} = \left[0,\ -\tfrac{kh}{m} + \tfrac{p}{m}\right]^T \tag{6.30}$$

$$\mathbf{C} = \left[1,\ 0\right]^T \tag{6.31}$$

$$\mathbf{q} = \left[q,\ \dot{q}\right]^T \tag{6.32}$$

where the time variable t is omitted for the simplicity of presentation.

Then, based on Eq. (6.26) along with a selected sampling time interval T, the discrete-time state-space model is derived:

$$\mathbf{q}_{k+1} = \mathbf{G}\mathbf{q}_k + \mathbf{H}u_k + \mathbf{P} \tag{6.33}$$

$$y_k = \mathbf{C}\mathbf{q}_k \tag{6.34}$$

where $\mathbf{q}_k = \mathbf{q}(kT)$ and

$$\mathbf{G} = e^{\mathbf{A}T} \tag{6.35}$$

$$\mathbf{H} = \int_0^T e^{\mathbf{A}_k \tau} d\tau \mathbf{B} \tag{6.36}$$

To facilitate the motion control design, the reference state is denoted as $\mathbf{q}_d$ and the closed-loop system error is defined as $\mathbf{e}(kT) = \mathbf{q}_d - \mathbf{q}(kT)$. And then, in consideration of the sliding variable defined below:

$$s(k) = \mathbf{c}^T \mathbf{e}(k) \tag{6.37}$$

the system stability is guaranteed if the switching function is maintained as $s(k) = 0$. The sliding function is selected such that $\mathbf{c}^T \mathbf{H} \neq 0$ is held. The desired performance of closed-loop system is ensured by the vector $\mathbf{c}$, that can be determined by using a quadratic optimization approach [5].

Taking into account the discrete-time dynamics model (6.33), the perturbed system is resulted:

$$\mathbf{q}_{k+1} = \mathbf{G}\mathbf{q}_k + \Delta\mathbf{G}[\mathbf{q}(k)] + \mathbf{H}u_k + \mathbf{P} \tag{6.38}$$

where $\Delta\mathbf{G}[\mathbf{q}(k)]$ is a bounded nonlinear function, which represents the model uncertainties and incorrectness.

Afterward, a nonswitching type of reaching law, that is appropriate for the perturbed system (6.38), is designed as follows.

$$s(k+1) = [1 - f(s(k))]s(k) - \widetilde{S}(k) - \widetilde{P}(k) + S_1 + P_1 \tag{6.39}$$

where S_1 and P_1 are the mean values of $\widetilde{S}$ and $\widetilde{P}$, respectively. $\widetilde{S}(k) = \widetilde{S}[(\mathbf{q}_k)] = \mathbf{c}^T \Delta\mathbf{G}[\mathbf{q}(k)]$ represents the lumped effect of the model uncertainty. $\widetilde{P}(k) = \mathbf{c}^T \mathbf{P}(k)$ denotes the external disturbance. Both $\widetilde{S}(k)$ and $\widetilde{P}(k)$ are unknown. Besides, f is a constant factor which is defined as a function $f(s)$: $\mathfrak{R} \rightarrow (0, 1]$ as follows.

$$f(s(k)) = \frac{s_0}{|s(k)| + s_0} \tag{6.40}$$

where $s_0 > S_2 + P_2$ is a constant that allows the generation of a proper compromise between the magnitude of the input generated by the presented controller and the robustness of the system designed in accordance with the given reaching law.

Define S_U and S_L as the upper and lower bounds for $\widetilde{S}$, and P_U and P_L as the upper and lower bounds for $\widetilde{P}$, respectively, as follows.

$$S_L \le \widetilde{S} \le S_U, \quad P_L \le \widetilde{P} \le P_U. \tag{6.41}$$

In addition, define S_2 and P_2 as the maximum admissible deviations of $\widetilde{S}$ and $\widetilde{P}$, respectively, that are given as follows.

$$S_1 = (S_U + S_L)/2, \quad P_1 = (P_U + P_L)/2 \tag{6.42a}$$

$$S_2 = (S_U - S_L)/2, \quad P_2 = (P_U - P_L)/2 \tag{6.42b}$$

According to Eqs. (6.37)–(6.39), the control signal $u(k)$ can be solved as follows.

$$u(k) = -(c^T H)^{-1}\{[1 - f(s(k))]s(k) + \mathbf{c}^T \mathbf{G}q(k) - \mathbf{c}^T \mathbf{x}_d + S_1 + P_1\} \tag{6.43}$$

6.4.2 *Stability Analysis*

Based on the designed reaching law (6.39), the stability of the given system is analyzed in this section.

Substituting Eq. (6.40) into Eq. (6.39), the nonswitching type of reaching law is expressed as:

$$s(k+1) = \frac{|s(k)|s(k)}{|s(k)| + s_0} - \widetilde{S}(k) - \widetilde{P}(k) + S_1 + P_1 \tag{6.44}$$

We can observe from Eq. (6.44) that $s(k+1)$ increases as $s(k)$ and $S_1 + P_1 - \widetilde{S}(k) - \widetilde{P}(k)$ increase. Additionally, referring to Eqs. (6.41) and (6.42a), the following inequality can be derived given any $j \ge 0$.

$$|S_1 + P_1 - [\widetilde{S}(j) + \widetilde{P}(j)]| \le S_2 + P_2 \tag{6.45}$$

Thus, if for some moment k, the following relation satisfies:

$$|s(k)| \le \frac{s_0(S_2 + P_2)}{s_0 - (S_2 + P_2)} \tag{6.46}$$

we can obtain that

$$
\begin{aligned}
|s(k+1)| &\leq \frac{s_0^2(S_2+P_2)^2/[s_0-(S_2+P_2)^2]}{s_0(S_2+P_2)/[s_0-(S_2+P_2)]+s_0}+(S_2+P_2) \\
&= \frac{(S_2+P_2)^2}{s_0-(S_2+P_2)}+(S_2+P_2) \\
&= \frac{s_0(S_2+P_2)}{s_0-(S_2+P_2)}
\end{aligned}
\tag{6.47}
$$

Therefore, from Eqs. (6.46) and (6.47), we can conclude that if $s(k)$ meets the inequality condition (6.46), then $s(k+1)$ also satisfies the inequality. Additionally, the control law (6.43) will drive the system trajectory to a subset Ω asymptotically, that is defined below:

$$
\Omega = \{x : |s(x)| \leq \frac{s_0(S_2+P_2)}{s_0-(S_2+P_2)}\}
\tag{6.48}
$$

Detailed proof of the stability of the control scheme is presented in [13]. We can observe from Eq. (6.43) that the control action does not contain a discontinuous control item. Thus, one uniqueness of the introduced control scheme lies in the inherent chattering-free property. The performance of the gripper device with the developed control scheme is validated by carrying out experimental studies in the following sections.

6.5 Prototype Fabrication and Performance Testing

6.5.1 Prototype Fabrication

In this work, the designed constant-force microgripper is machined with a plate material of Al-6061 by the wire electrical discharge machining (EDM) approach. Figure 6.8 shows the photograph of the developed prototype of the constant-force microgripper. With the dimension of 81.5 mm × 70.2 mm × 4 mm, the gripper owns a compact size. In view of the force and stroke requirements, a PSA actuator (model: P-888.91, from Physik Instrumente Co., Ltd.) is chosen for the drive. The PSA provides a nominal travel stroke of 32 μm, that is actuated by a commercial high-voltage amplifier (model: EPA-104, from Piezo System, Inc.). To measure the output position of gripper tip, a capacitive displacement sensor (model: D-510.050, from Physik Instrumente Co., Ltd.) is employed. Additionally, the force is detected with a force sensor (model: LSB200, from FUTEK Advanced Sensor Technology Inc.), that provides a measurement range of 25 lb.

To calibrate the capacitive position sensor, a laser displacement sensor (model: LK-H055, from Keyence Corporation) is adopted, that provides a resolution of 25 nm

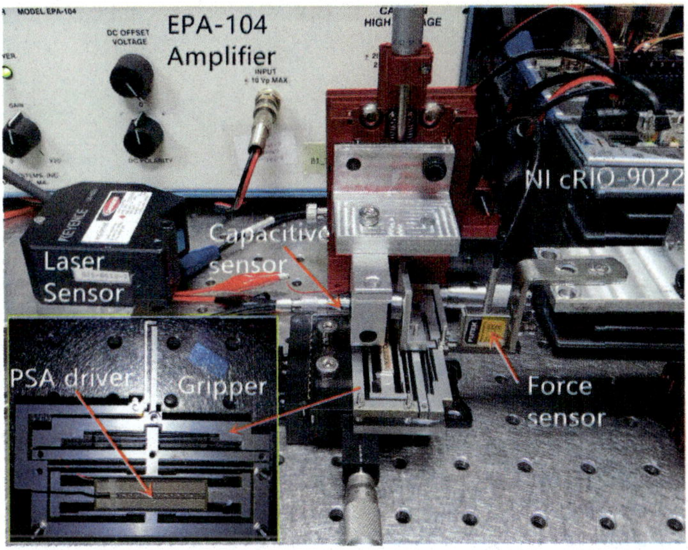

Fig. 6.8 Experimental setup of the constant-force microgripper

within the measuring range of 20 mm. Moreover, a real-time (RT) controller (model: cRIO-9022, from National Instruments (NI) Corp.) combined with reconfigurable chassis (model: cRIO-9118, from National Instruments Corp.) is adopted to implement the control algorithm. The I/O modules (model: NI-9263, NI-9237, and NI-9215, from National Instruments Corp.) are used for producing analog driving signals and acquiring the position and force sensor output signals. The cRIO-9118 chassis comprises a field-programmable gate array (FPGA) core, and the associated cRIO-9022 RT controller is connected with a personal computer via an Ethernet port. The control algorithms are programmed with NI LabVIEW software to implement a deterministic RT control of the gripper device.

6.5.2 Gripping Range and Hysteresis Tests

Using the given position sensor, the gripping range for the gripper is investigated experimentally. The experimental result of the gripping range test is shown in Fig. 6.9, which is obtained by applying a sinusoid signal with the frequency of 0.5 Hz and amplitude of 10 V to the PSA. We can observe that the maximum displacement for the actuated gripper tip is about 220 μm. Thus, the gripper possesses a gripping range of 220 μm. In comparison with the simulation result for the gripping range, the experimental result is 10% lower. The discrepancy is mainly induced by the manufacturing error of the gripper and preloading effect of the PSA actuator, which are not considered in the simulation study.

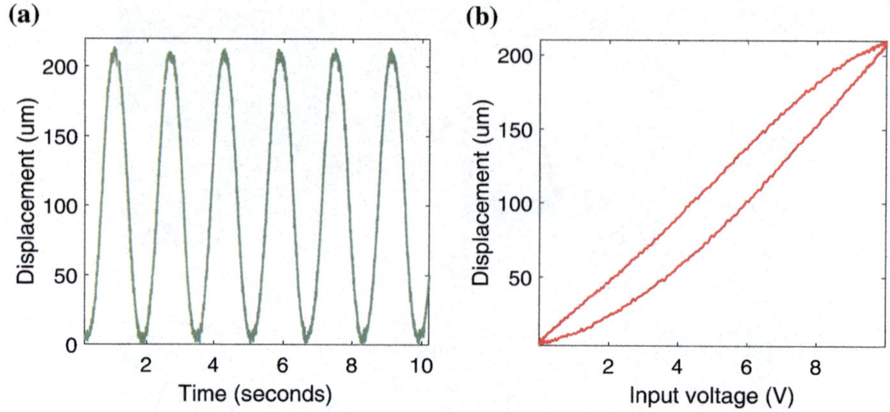

Fig. 6.9 Experimental results of gripping range by applying a 0.5 Hz, 10-V amplitude sinusoidal input: **a** time history results and **b** hysteresis curves

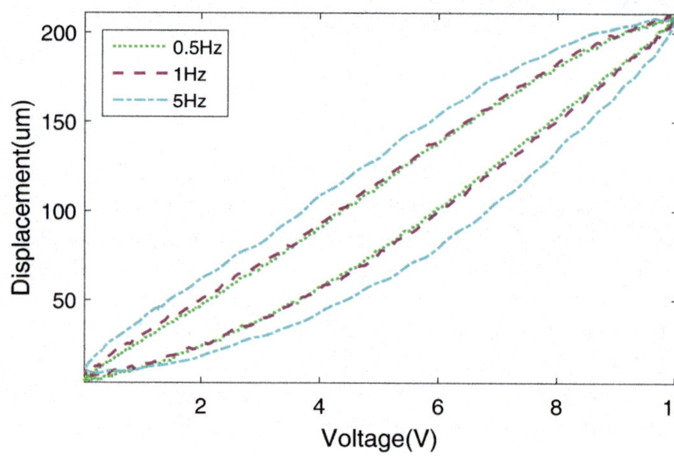

Fig. 6.10 Hysteresis loops obtained by applying sinusoidal voltage inputs with different frequencies

To test the hysteresis property for the piezo-driven microgripper, sinusoidal waves with the frequencies of 0.5, 1, and 5 Hz are adopted for the open-loop testing of the gripper. Figure 6.10 depicts the experimental results. We can see that the shape of the hysteresis loop is dependent on two factors in terms of the amplitude and frequency for the input signal. As expected, as the amplitude and frequency for the input signal increase, the width of the hysteresis loop increases. Thus, to achieve a precision motion control under the inputs with high frequency and large amplitude, it is indispensable to suppress the hysteresis effect.

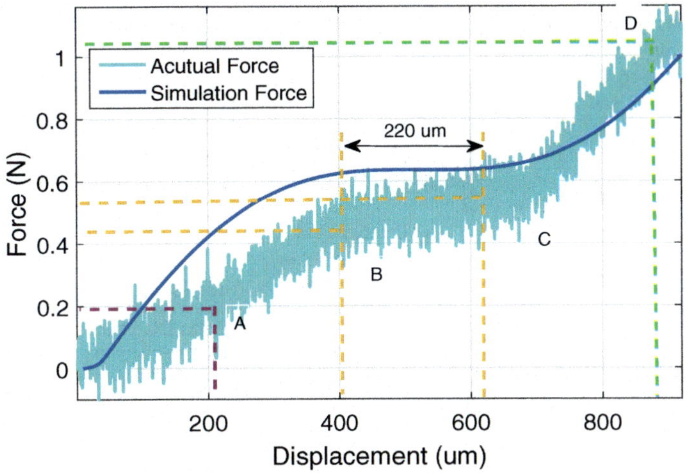

Fig. 6.11 Simulation and experimental results of force–displacement relationship for the gripper

6.5.3 Force–Displacement Relation Test

The force–displacement relation of the fabricated gripper is obtained by driving the gripper and measuring the corresponding displacement and force. Figure 6.11 shows the experimental result. The simulation result is also given in Fig. 6.11 for a clear comparison. We can see that the experimental result is in good consistency with the simulation result with slight discrepancy. The discrepancy is attributed to the fabrication errors for the gripper parameters. Experimental results demonstrate that the gripper provides a constant force output within the motion range around 220 μm. The test results confirm the fine performance of the developed constant-force microgripper.

6.5.4 Dynamics Performance Test

The dynamics performance of the microgripper is examined by the frequency response approach. Specifically, a swept sine wave with the amplitude of 0.1 V and the frequency of 1–500 Hz is adopted to actuate the PSA via the high-voltage amplifier. The position response is detected by using the capacitive displacement sensor.

To investigate the dynamics characteristics at different operating points, the frequency responses are generated at four typical points (annotated by A, B, C, and D in Fig. 6.11). In the experimental study, the gripper is driven to attain every working point initially by applying a DC voltage. And then, the frequency response data are obtained by applying the swept sine signal. Figure 6.12 shows the experimental results. We can observe that the resonant frequency decreases from the operating

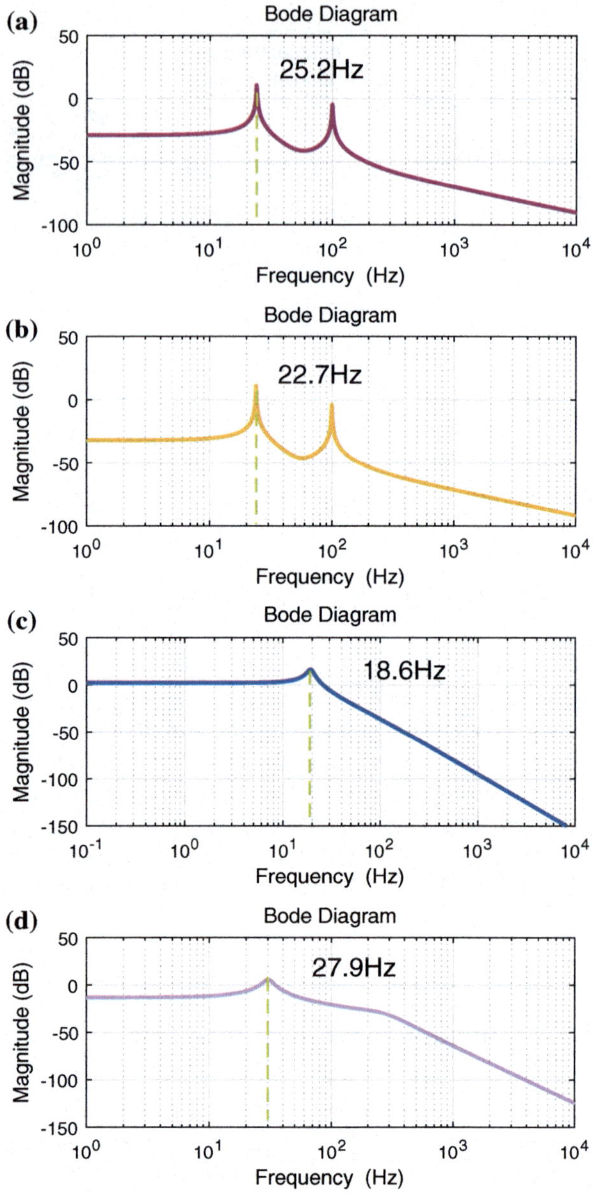

Fig. 6.12 Magnitude plots of Bode diagrams for the gripper device. **a–d** show the frequency responses at the points A, B, C, and D, respectively

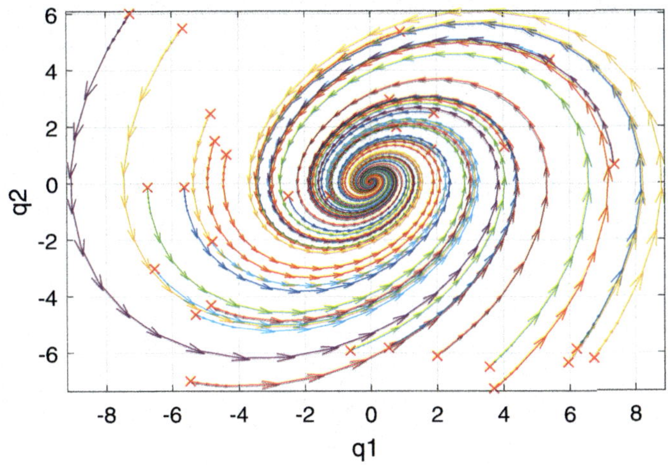

Fig. 6.13 Velocity vector field of the gripper system. q_1 and q_2 denote the velocity and acceleration of the system, respectively

point A to C and then increases from point C and D gradually. That is, the resonant frequency attains the minimal value of 18.6 Hz around point C, that is located within the constant-force range. In theory, the zero stiffness indicates a zero resonant frequency for the mechanism. The presence of a low resonant frequency (at 18.6 Hz) is induced by the unbalanced stiffness for the gripper device.

According to the experimental data, a second-order transfer function model of the plant is identified. The vector field diagram of the gripper system is shown in Fig. 6.13, which is obtained based on the transfer function model at the operating point C in Fig. 6.11. With random initial points and 10 unit time spans, we can observe that there is only one equilibrium point in the system and the state converges to the original point (0, 0) as $t \to \infty$. Thus, the gripper system plant is stable with a stable limit cycle.

6.6 Closed-Loop Experimental Studies

In this section, the closed-loop experimental studies are conducted with the proposed control law. Figure 6.14 depicts the block diagram of the motion control for the gripper system. It is noted that the force sensor is only adopted to detect the force signal. The parameters for the dynamics model (6.25) of the gripper can be identified by system identification approach. For example, by adopting a sampling time interval of 0.001 s, the parameters for the second-order dynamics model (at the operating point C) are given below.

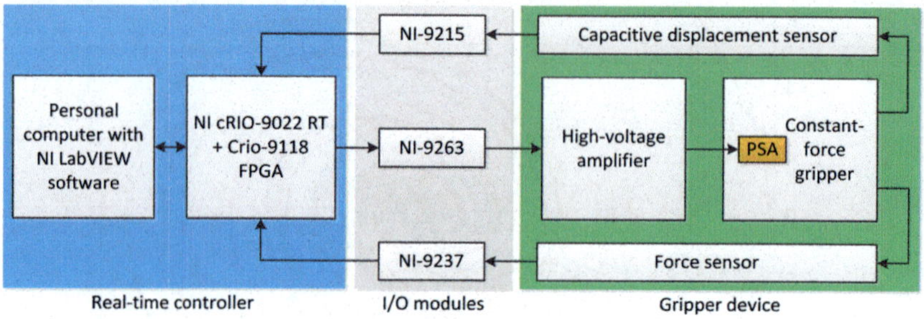

Fig. 6.14 Block diagram for the gripper control system

$$G = \begin{bmatrix} 0.04258 & -12.31 \\ 11.84 & -6.444 \end{bmatrix} \tag{6.49}$$

$$H = \begin{bmatrix} 0.0432 \\ -0.6407 \end{bmatrix} \tag{6.50}$$

$$C = [1,\ 0] \tag{6.51}$$

The vector **c** of sliding function parameter is chosen as follows.

$$\mathbf{c} = [1,\ 0.675] \tag{6.52}$$

Additionally, in consideration of Eq. (6.52) and the disturbance range $p(k) < 0.02$, $P_1 = 0$ and $P_2 = 0.02$ are chosen. Afterward, by selecting $S_1 = 0$ and $S_2 = 0.5$ and taking into account the stability requirement $s_0 > S_1 + S_2$, we can derive that $s_0 > 0.52$.

6.6.1 Resolution Testing Result

Firstly, an experimental study is carried out to test the resolution of the gripper system. Specifically, a consecutive step signal is applied to examine the minimum resolution of gripper system with the designed controller. In particular, the experiment is performed using a 20-nm step size. In the experiment, the displacement increases to 100 nm and then decreases to the home position. The experimental result is shown in Fig. 6.15.

We can observe that each step with the amplitude of 20 nm is precisely tracked and clearly identified. The maximum tracking error is around 10 nm, that is close to the resolution of the adopted capacitive displacement sensor. The experimental result indicates that the gripper system with the proposed controller is able to provide a better than 20 nm positioning resolution.

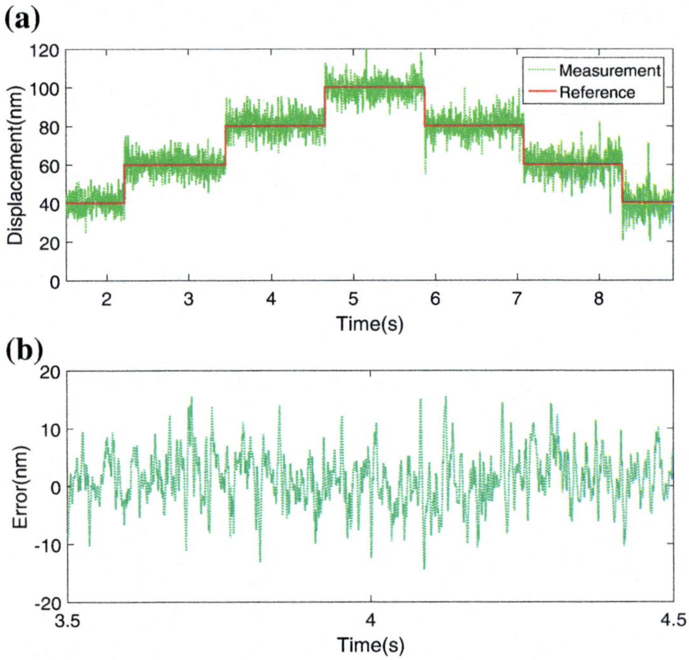

Fig. 6.15 Resolution testing results using 20-nm consecutive step input. **a** Measurement output and reference input; **b** tracking errors

6.6.2 Grasp-Hold-Release Operation Testing Result

When gripping an object, the superiority of a constant-force gripper allows the generation of constant grasp force by using a position control alone. It ensures the safety of the grasped object without adopting a force sensor and controller. In order to validate this feature of the constant-force gripper, the grasp test of a microwire of copper is performed. Figure 6.16a–d exhibits the position tracking result, controller output, position error, and force signal, respectively. We can observe that three steps are utilized to grasp a bundle of copper wires of 25-μm diameter. Firstly, the gripper jaw closes with a constant velocity of $v_r = 40\mu m$ to grasp the copper wire. Afterward, the gripper holds the copper wire for a time period of 1.5 s. Next, the gripper releases the copper wire with the constant velocity of v_r.

During the gripping procedure, the maximal position error of around 1 μm and root-mean-squared error (RMSE) of 0.247 μm are generated, as shown in Fig. 6.16c. It reveals that the designed sliding mode controller satisfies the accuracy requirement of gripping. Figure 6.16d indicates that although the position varies from initial value of 20 μm to the maximum value of 60 μm, the force remains unchanged at 530 mN. The experimental results confirm the efficiency of developed constant-force gripper.

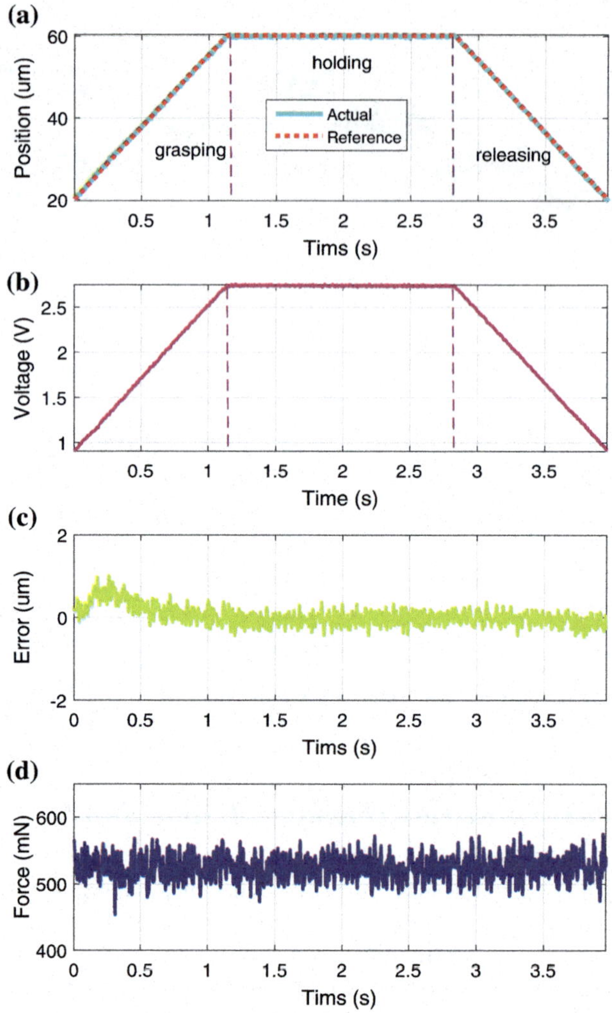

Fig. 6.16 Experimental results for gripping a bundle of copper wire. **a** Position control result; **b** control action; **c** position control error; **d** output force signal

Specifically, the grasp force is kept constant in the working range despite the position variation. Furthermore, Fig. 6.16b illustrates that no chattering phenomenon is presented in the control action, that is one of the advantages of the adopted control scheme.

Additionally, Fig. 6.17 shows the snapshots for the grasp procedure. It exhibits the closing, contact, and releasing stages of the manipulation, respectively.

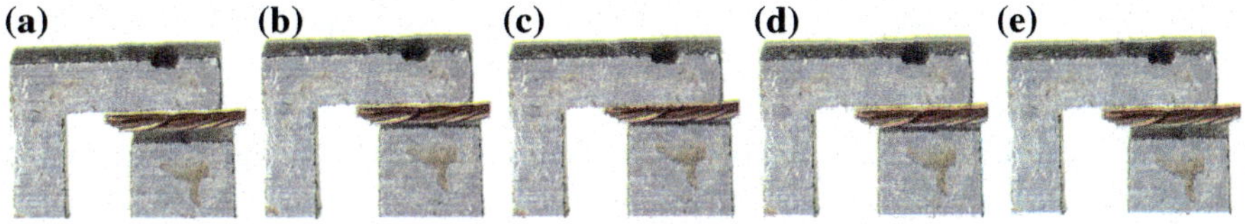

Fig. 6.17 Snapshots of gripping procedure for a bundle of copper wire. (a)–(b) grasping, (c) holding, and (d)–(e) releasing manipulations

6.6.3 Further Discussion

Table 6.2 shows the comparison of the gripper performances which are evaluated by the experimental and simulation studies. We can observe that the experiment result of the constant output force is 530 mN with a deviation of 14.5% from the desired value. The discrepancy is mainly attributed to the machining tolerance for the gripper prototype. In particular, we find that the result is rather sensitive to the variation of the initial inclined angle for the bistable beams. In the future, the gripper prototype will be fabricated more precisely to generate a better performance.

To illustrate the advantage of the developed constant-force gripper, the control results are compared to those of a conventional compliant gripper with positive stiffness [18]. Regarding the constant-force gripper, the stand-alone position control is able to cater for the requirement of safe grasp-hold-release manipulation, that reduces the system complexity efficiently. By contrast, the conventional positive-stiffness gripper demands a complex control strategy to realize the regulation of both position and force values. For comparison purpose, the experimental results for the two implemented control schemes are shown in Table 6.3. The advantage of constant-force gripper is obvious to see. Specifically, the use of the constant-force gripper not only eliminates the adoption of force control strategy, but also enables a better performance for the position control. In particular, it generates an RMSE of

Table 6.2 Comparison of the results of constant-force gripper

Performance	Experiment result	Designed value	Error (%)
Constant force output	530 mN	620 mN	14.5
Actuation module output	220 μm	246 μm	10.5

Table 6.3 Performance comparison of two different grippers

Gripper type	Position RMSE (μm)	Force RMSE	Overall time (s)
Conventional gripper	0.202	0.747 mN	5.29
Constant-force gripper	0.141	–	3.96

0.141 μm. In comparison with the conventional gripper with force/position switching control, the position error has been reduced by over 30%.

As for the constant-force gripper device, the elimination of force and position switching control allows the complexity reduction of control design. Moreover, it enables the improvement for the operation speed. Table 6.3 indicates that the conventional gripper requires the overall operation time of 5.29 s to realize a grasp-hold-release manipulation, while the constant-force gripper only needs 3.96 s to execute the same manipulation. Indeed, the time of 3.96 s is not the limit of the developed constant-force gripper. In accordance with the features and sizes of different objects, the grasping and releasing velocities can be properly adjusted to generate different objectives. With the conventional gripper, there is a time lag between the position and force switching, that induces a relative large control error.

As the constant-force value is governed by the mechanical design parameters of the gripper structure, the output force of constant-force gripper cannot be arbitrarily adjusted. In order to change the constant-force magnitude of the gripper, the parameters of the constant-force module (including the inclination angle, length, and width of the beam) can be redesigned. Besides, the future work will be conducted toward the design of a constant-force gripper with adjustable output force.

6.7 Conclusion

The design and control of a compliant constant-force gripper based on buckled fixed-guided beam is proposed in this chapter. The mechanism design and sliding mode motion control design are presented in detail. Experimental results of the gripping range testing show that the gripper is able to provide a constant-force of 530 mN with a motion range of 220 μm. The experimental results of grasp-hold-release operation of micro-object demonstrate that the proposed gripper has better performance and faster operation speed in comparison with conventional gripper.

References

1. Al Janaideh, M., Rakheja, S., Su, C.Y.: Experimental characterization and modeling of rate-dependent hysteresis of a piezoceramic actuator. Mechatronics **19**(5), 656–670 (2009)
2. Boukari, A.F., Carmona, J.C., Moraru, G., Malburet, F., Chaaba, A., Douimi, M.: Piezo-actuators modeling for smart applications. Mechatronics **21**(1), 339–349 (2011)
3. Clayton, G.M., Tien, S., Leang, K.K., Zou, Q., Devasia, S.: A review of feedforward control approaches in nanopositioning for high-speed SPM. J. Dyn. Sys., Meas. Control **131**(6), 061101–1–061101–19 (2009)
4. Holst, G.L., Teichert, G.H., Jensen, B.D.: Modeling and experiments of buckling modes and deflection of fixed-guided beams in compliant mechanisms. J. Mech. Des. **133**(5), 051002 (2011)
5. Janardhanan, S., Kariwala, V.: Multirate-output-feedback-based lq-optimal discrete-time sliding mode control. IEEE Trans. Autom. Control **53**(1), 367–373 (2008)

6. Kim, K., Liu, X., Zhang, Y., Sun, Y.: Nanonewton force-controlled manipulation of biological cells using a monolithic MEMS microgripper with two-axis force feedback. J. Micromech. Microeng. **18**(5), 055013 (2008)
7. Komati, B., Clevy, C., Lutz, P.: High bandwidth microgripper with integrated force sensors and position estimation for the grasp of multi-stiffness microcomponents. IEEE/ASME Trans. Mechatron. **16**(1), 141–150 (2016)
8. Lamers, A.J., Sanchez, J.A.G., Herder, J.L.: Design of a statically balanced fully compliant grasper. Mech. Mach. Theory **92**, 230–239 (2015)
9. Lan, C.C., Lin, C.M., Fan, C.H.: A self-sensing microgripper module with wide handling ranges. IEEE/ASME Trans. Mechatron. **16**(1), 141–150 (2011)
10. Liaw, H.C., Shirinzadeh, B.: Robust generalised impedance control of piezo-actuated flexure-based four-bar mechanisms for micro/nano manipulation. Sens. Actuatat. A Phys. **148**(2), 443–453 (2008)
11. Lin, C.J., Chen, S.Y.: Evolutionary algorithm based feedforward control for contouring of a biaxial piezo-actuated stage. Mechatronics **19**(6), 829–839 (2009)
12. Liu, F., Pang, X., Liu, Z., Zhang, T., Huang, H., Zhu, J.: Negative-stiffness vibration isolation. In: Proceedings of the 3rd International Nanoelectronics Conference (INEC), pp. 154–155 (2010)
13. Liu, Y., Zhang, Y., Xu, Q.: Design and control of a novel compliant constant-force gripper based on buckled fixed-guided beams. IEEE/ASME Trans. Mechatron. **22**(1), 476–486 (2017)
14. Rakotondrabe, M., Ivan, I.A.: Development and force/position control of a new hybrid thermo-piezoelectric microgripper dedicated to micromanipulation tasks. IEEE Trans. Autom. Sci. Eng. **8**(4), 824–834 (2011)
15. Tolou, N., Pluimers, P., Jensen, B.D., Magleby, S., Howell, L.L., Herder, J.L.: Near-zero-stiffness linear motion stage with high orthogonal and out-of-plane stiffness. In: Proceedings of 1st DSPE Conference on Precision Mechatronics, pp. 1–2. Deurne, The Netherlands (2012)
16. Wang, J.Y., Lan, C.C.: A constant-force compliant gripper for handling objects of various sizes. J. Mech. Des. **136**(7), 071008 (2014)
17. Xu, Q.: Identification and compensation of piezoelectric hysteresis without modeling hysteresis inverse. IEEE Trans. Ind. Electron. **60**(9), 3927–3937 (2013)
18. Xu, Q.: Design and smooth position/force switching control of a miniature gripper for automated microhandling. IEEE Trans. Ind. Inform. **10**(2), 1023–1032 (2014)
19. Xu, Q.: Design and development of a novel compliant gripper with integrated position and grasping/interaction force sensing. IEEE Trans. Autom. Sci. Eng. **14**(3), 1415–1428 (2017)
20. Zhang, J., Merced, E., Seplveda, N., Tan, X.: Optimal compression of generalized Prandtl-Ishlinskii hysteresis models. Automatica **57**, 170–179 (2015)
21. Zhao, J., Jia, J., He, X., Wang, H.: Post-buckling and snap-through behavior of inclined slender beams. J. Appl. Mech. **75**(4), 041020 (2008)
22. Zhou, Q., Korhonen, P., Laitinen, J., Sjövall, S.: Automatic dextrous microhandling based on a 6-DOF microgripper. J. Micromechatron. **3**(3), 359–387 (2006)

Chapter 7
Design and Development of a Flexure-Based Compact Constant-Force Robotic Gripper

Abstract This chapter presents the design of a novel flexure-based compliant gripper with constant gripping force and compact structure size for cell micromanipulation applications. The gripper removes the use of force sensor and provides a near constant force output via its mechanical structure, which greatly simplifies the system design process. The compact size of the gripper is achieved by the serial connection of a bistable beam and a positive-stiffness beam. Moreover, a combined mechanism, which can alter the fixing angle of the two gripper jaws, is developed to enlarge the handling size. Analytical modeling and finite element analysis are conducted to predict the gripper performance. A prototype gripper is fabricated by 3D printer, and a series of experiments are carried out to verify its performance. Grasp testing of crab egg embryos has been carried out to demonstrate its effectiveness in biological micromanipulation application.

7.1 Introduction

Robotic gripper (or grasper, tweezer) is a device which can transmit the input displacement into output gripping motion. It is usually adopted as an end-effector of robots to pick up or transport objects [4, 11]. To meet the requirement of micro-objects operation, various compliant grippers have been developed [30, 34]. As compared with traditional rigid-body grippers, compliant grippers exhibit many advantages, such as no wear, no backlash, and no friction [17]. In addition, compliant mechanisms can be easily miniaturized into micro/nanoscale [10]. As compliant grippers are adopted in micromanipulation systems, it is a challenging task to produce a suitable gripping force. Different from macro-objects, micro-objects are more sensitive to the change of contact force [29]. The reason lies in that micro-objects are prone to be destroyed without regulating the handling force.

Conventionally, the output force of the gripper jaw can be measured by attaching a force sensor on the robotic gripper [1, 9, 21]. However, an extra space is needed to mount the sensor, which increases the gripper size and structure complexity. Because the safe contact force of a micro-object is usually very small, the satisfactory sensor requires a complicated processing technique [18]. Moreover, using

Q. Xu, *Micromachines for Biological Micromanipulation*,
https://doi.org/10.1007/978-3-319-74621-0_7

a sensor-controller system to regulate the gripping force needs a complex control algorithm design [8, 31]. As another approach, computer vision technology can be employed to detect and control the gripping force [23]. By measuring the deformation of the manipulated objects, the gripping force can be estimated based on an elastic model. However, the contact-induced deformation is very small, and such method is rarely practicable [27]. Moreover, both methods require a skilled operator or a precision actuator, which imposes a high cost on hardware and operation. In addition, variable-compliance mechanism has been introduced to adjust the robustness of an underactuated grasper against external force perturbations [24].

Alternatively, the development of constant-force mechanism (CFM) paves a new way to regulate the gripping force [12, 26]. Unlike conventional elastic structures, the CFMs do not obey Hooke's law [19]. The stiffness of CFMs will become zero in their specific range of deformation. Actually, the essence of gripping force control is the overload protection [22]. By using CFM to drive the gripper jaw, the output force can be regulated as constant for a specific range of input displacement. As compared with traditional ways, the advantage of CFM is significant as it is capable of regulating a constant output force by eliminating the use of precision actuator and force sensor/controller. The removal of force sensor can reduce the cost and improve the gripper's compactness. Without using a force feedback control, the control algorithm design can be greatly simplified. A simple step input signal can be used to drive the gripper. Moreover, the fact that the constant output force is produced in a specific range of displacement alleviates the requirement on the actuator's precision.

In the literature, a few of constant-force grippers have been developed [2, 6, 14, 16]. However, the existing works exhibit some limitations. For instance, each gripper arm must be actuated independently. Thus, the gripper usually needs two sets of actuators and CFMs, which leads to a relatively large structure size and complicated operation process. To solve this issue, a new design is reported in recent work [13]. The working principle of the gripper is similar to that of former designs, while only one gripper arm is actuated. The design greatly reduces the gripper size and simplifies the operation process. However, the gripping motion is unidirectional and the one-arm-fixed strategy cuts its gripping stroke by half, which results in a small gripping range. It is not suitable for the grasp of micro-objects with wide distribution of sizes. Another constant-force compliant gripper has been presented in [27], which can handle objects of various sizes. However, the output force is too large (in Newton level) to be adopted for micro-object grasp. This design cannot provide a suitable level of output force without sacrificing the compact structure size. In addition, to grasp micro-objects with various sizes, it is desirable to design a constant-force gripper with adjustable handling size. Nevertheless, no such design has been reported in the literature.

In this chapter, a new compliant constant-force gripper with both compact structure size and large gripping range is proposed, which can handle the objects of various sizes. A compact structure is necessary for operation in limited space for applications such as micromanipulation. Moreover, it provides a suitable force magnitude for the manipulation of micro-objects.

7.2 Mechanism Design

In this section, the mechanism design of a flexure-based gripper with constant gripping force, large gripping stroke, and adjustable handling size is presented.

The constant output force is obtained by using a CFM with zero stiffness. The zero-stiffness mechanism can be obtained by different ways, such as building block method and optimization approach. The former is a flexible approach by a proper connection of negative- and positive-stiffness mechanisms. In the literature, both parallel [12, 28, 33] and serial [25] connection schemes of negative- and positive-structures have been reported. The parallel connection scheme realizes a constant-force output intuitively. However, it induces an extra stress which increases the output force unnecessarily. As a result, long and thin beams have to be used to generate a suitable magnitude of output force, which enlarges the physical size of the structure. On the contrary, the serial connection scheme has a potential to produce a reasonable force magnitude with a compact size of the constant-force mechanism. Nevertheless, relatively limited work on serial connection scheme has been conducted. To the knowledge of the authors, only one constant-force mechanism has been reported with serial connection scheme in recent work [25]. No effort has been attempted to design a constant-force gripper using serial connection of negative- and positive-mechanisms.

7.2.1 Design of Constant-Force Module

Conventionally, the zero-stiffness structure is constructed by the parallel connection of the negative- and positive-stiffness structures, as shown in Fig. 7.1a. Such layout realizes constant-force output easily, but it leads to some problems. For instance, parallel connection will cause an extra stress which makes the output force increase greatly. To generate a suitable output force magnitude, long and thin beams are needed, which increases the size of the structure significantly. To overcome this issue, a serial connection scheme, as illustrated in Fig. 7.1b, is adopted in this work. It can eliminate the effect of extra stress and offer a reasonable force with a compact size. Giving the same output force value, the serial connection scheme is more compact than the parallel one.

As for a constant-force mechanism with parallel connection scheme, the positive- and negative-stiffness structures can be modeled separately. While in the serial scheme, the structures should be considered at the same time. Regarding the modeling approach of constant-force mechanism, the popular pseudo-rigid-body model (PRBM) [20] and elliptic integral model [7] are relatively complex and model errors always exist. By contrast, finite element analysis (FEA) provides a more convenient and accurate way to predict its performance. Hence, FEA is adopted to evaluate the performance of the designed constant-force mechanism in this work.

In particular, the bistable beam is the core part of CFM. The beam will exhibit negative stiffness in forced deformation owing to its buckling property. Hence, in the design procedure, the bistable beam is designed first. Afterward, a leaf flexure (acting

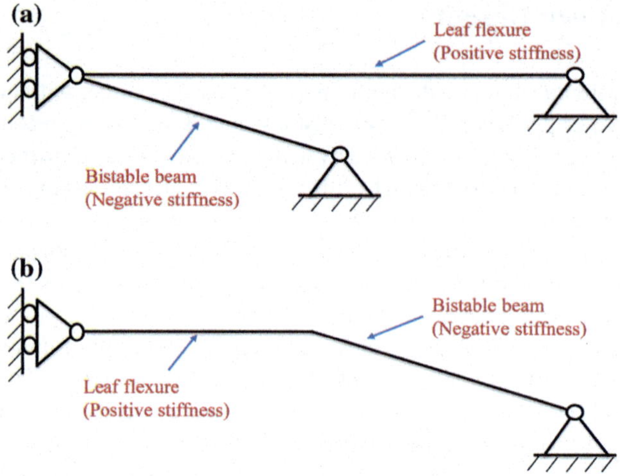

Fig. 7.1 Sketch diagram of zero-stiffness mechanism with **a** parallel connection scheme and **b** serial connection scheme

as positive-stiffness mechanism) is connected to the one end of the bistable beam. The beam parameters are then adjusted until the whole structure exhibits zero-stiffness property with large deformation range.

7.2.2 Design of Gripper Jaw Module

Gripper jaws are the gripper's end-effector, which contacts the manipulated micro-objects directly. The working principle of the gripper jaws is shown in Fig. 7.2a. To generate a large gripping range, a displacement amplifier is required to amplify the gripping stroke. There are two main methods to amplify the displacement, i.e., triangular amplification principle [3] and lever amplification principle [5]. Unlike the former, the latter one possesses a simple mechanism and it is adopted in this work. Moreover, as the selected material (i.e., ABS plus) is soft, the triangular-type amplifier will suffer from unwanted lost motion. On the contrary, the lever-type amplifier can avoid it perfectly as it has a higher efficiency. Thus, a lever-type amplifier is chosen to magnify the gripping stroke.

The schematic diagram of its working principle is shown in Fig. 7.2b, where Δx and Δy represent the input and output displacements, respectively. The increment of input Δx will cause an increment of output Δy by the lever arm. Considering the relationship of similar triangles, we can get the following relationship:

$$\frac{L - \Delta x}{\Delta y} = \frac{d}{\Delta x} \tag{7.1}$$

where L and d denote the length and width of the gripper jaw, respectively.

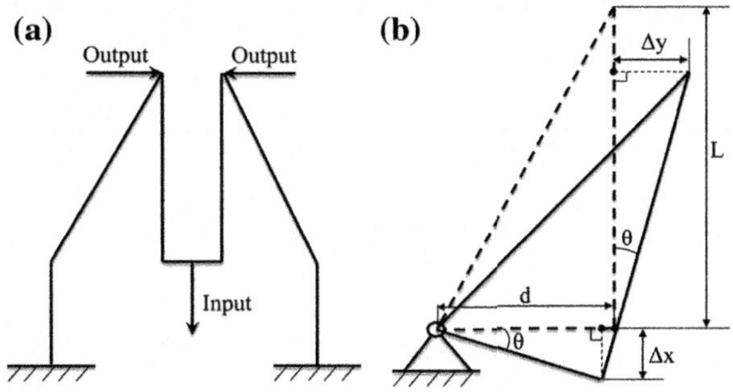

Fig. 7.2 Illustrations of **a** gripping jaws and **b** working principle of the gripping jaw

From Eq. (7.1), the following equation can be obtained:

$$\Delta x \cdot L - \Delta x^2 = d \cdot \Delta y \tag{7.2}$$

Considering that the input displacement Δx is very small as compared with the length and width of the jaw, the squared value (Δx^2) of the input displacement is ignored. Then, the equation of the amplification ratio can be simplified as:

$$\beta_0 = \frac{\Delta y}{\Delta x} \approx \frac{L}{d} \tag{7.3}$$

As the gripper jaws are driven bidirectionally, the amplification ratio is twice the one with unidirectional drive. Thus, the amplification ratio can be derived as follows.

$$\beta = 2\beta_0 \approx \frac{2L}{d} \tag{7.4}$$

It is observed from Eq. (7.4) that the amplification ratio is governed by the length (L) and width (d) of the gripper jaws. Equation (7.4) can be used to design the parameters of the gripper jaws.

7.2.3 Design of the Gripper Layout

The entire constant-force gripper is composed of three components in terms of driving section, guiding section, and operating section, as illustrated in Fig. 7.3a. The driving and operating sections involve the linear actuator and jaws, respectively. The CFM is selected as the guiding section. The function of this part is to change the property of driving force and to transmit it into the gripper jaws. A CFM behaves like a force-limiting structure in the gripper. To realize this function, the input and output sides of CFM are connected to the actuator and the jaws, respectively. The schematic diagram of the designed constant-force gripper is shown in Fig. 7.3b.

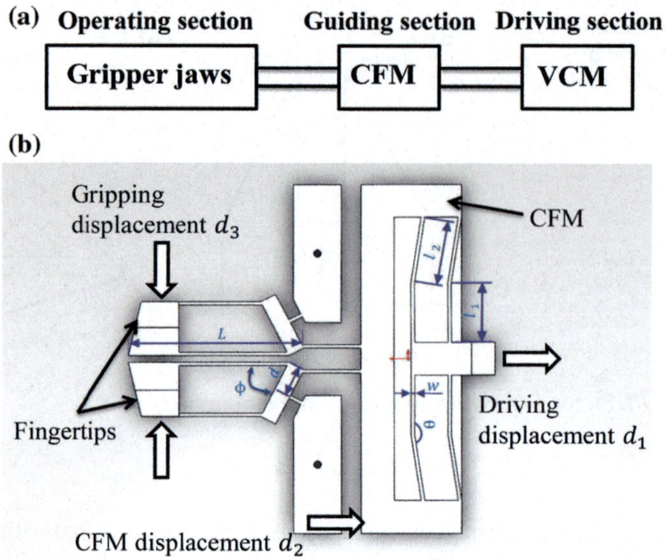

Fig. 7.3 **a** Schematic of a constant-force gripper; **b** CAD model of the designed gripper

7.2.3.1 Enhanced Supporting Structure

The CFM is designed as a symmetric structure to guarantee the direction of output force. To avoid rotational movement, at least two beams are used in parallel on one side. A necessary condition to produce a constant force of CFM is that one end of the bistable beams should be entirely fixed. However, due to the low stiffness of selected material (ABS plus), the use of ordinary supporting structure (see Fig. 7.4a) cannot meet the requirement on boundary constraint. Therefore, a stiffening platen is added at the bottom to enhance the strength of the supporting structure, as shown in Fig. 7.4b.

A comparison study between the ordinary supporting structure and enhanced supporting structure is conducted, and the simulation result is shown in Fig. 7.4c. The ideal output force property is used as a reference. It is seen that the ordinary supporting structure cannot produce a constant output force. In contrast, the enhanced supporting structure matches well with the ideal condition, which demonstrates the effectiveness of the proposed design of enhanced supporting structure.

7.2.3.2 Adjustable Gripper Jaw

To make the gripper adjustable for handling objects with various sizes, a device which can adjust the original position of the gripper jaws is needed. In the designed gripper, a device is introduced which can realize such function by preloading.

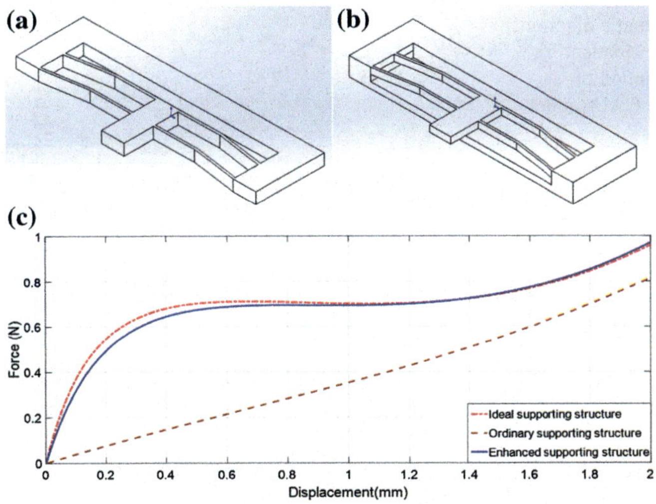

Fig. 7.4 Illustrations of **a** the ordinary supporting structure, **b** the proposed enhanced supporting structure, and **c** force–displacement relation of different structures

As shown in Fig. 7.5a, two spline holes are fabricated on the support structure, and two splines are fixed at the base. The preloading is realized by the assembly of gripper and base using splines. The gripping range is determined by the parameters of the gripper as follows.

$$\theta = 2\alpha \approx \arcsin(l_4/l_3) \tag{7.5}$$

where the angles α and θ are denoted in Fig. 7.5b.

By fixing the splines at the base, the original position of the gripper jaws can be adjusted discretely. The interval angle of adjustment is determined by the number of edges of the spline. In this case design, the splines with six edges are selected. The two splines can be adjusted with the same angle α to ensure that they are rotated simultaneously. It is notable that if the splines are permitted to rotate freely at the base before fixing, the jaw position of the gripper can be changed continuously by adjusting the installation angle α of the splines.

7.2.3.3 Assembly Scheme

A computer-aided design (CAD) model of developed gripper is shown in Fig. 7.5a. Two holes on the tips of jaws are used to mount the force sensor, and the convex part acts as the measurement target of laser displacement sensor. Note that the force and displacement sensors are used to calibrate the gripper's performance only. In practical application, the sensors are not needed and the structure can be further simplified. The holes in the middle part are used to fix the gripper onto the base. A linear actuator is connected to the gripper by the hole located on the end terminal of the gripper.

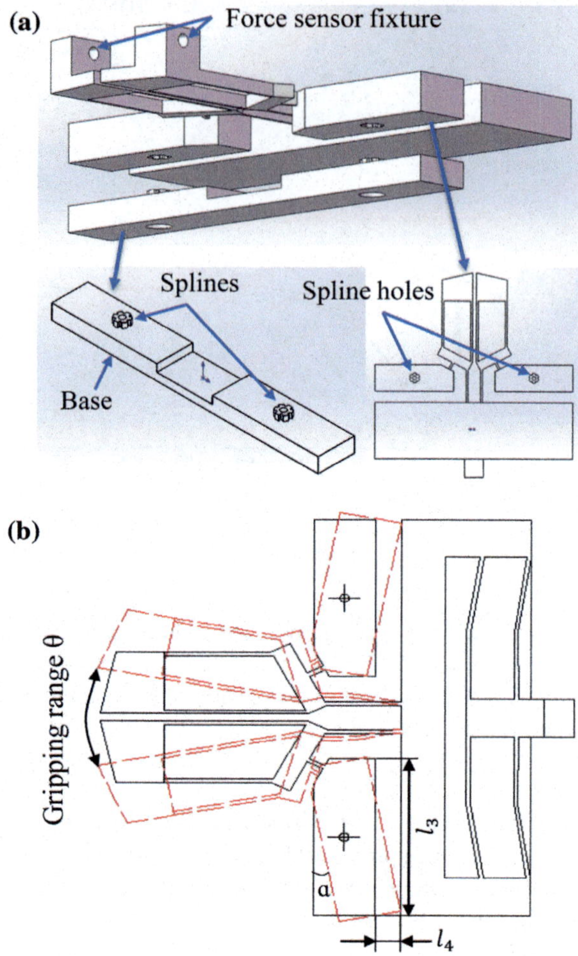

Fig. 7.5 Schematic diagram of adjusting mechanism for the original position of gripper jaws. **a** CAD model; **b** top view

7.3 Parametric Design

To develop a compliant gripper with desired gripping force, gripping stroke, and gripping range, the key parameters are carefully designed as outlined in this section. As a case study, the parameters are selected as shown in Table 7.1, where w donates the in-plane width of the beam and t represents the out-of-plane thickness of the beam. Other parameters are denoted in Fig. 7.3.

Table 7.1 Main structural parameters of the designed gripper

Component	Parameter	Value	Unit
CFM	l_1	18	mm
	l_2	20.25	mm
	w	1	mm
	t	3	mm
	θ	171	degree
Jaw	L	51.88	mm
	d	9	mm
	b	5	mm
	ϕ	62.7	degree

7.3.1 Actuation Force Consideration

The constant force is generated by the beams' deformation. The value of the reaction force should be assessed to ensure that the selected actuator is able to drive it to produce a reasonable stroke. Substituting the designed parameters into FEA model, the reaction force is obtained as shown in Fig. 7.6a.

It is observed that the actuation force of the CFM is about 0.7 N. Ranging from 0 to 0.8 mm, the force increases gradually. This phase is related to the gripping stroke. The constant-force stroke ranges from 0.8 to 2.4 mm, which is referred as overload protection stroke. The displacement range beyond 2.4 mm is called overload stroke in this work. Note that the displacement shown in the lateral axis in Fig. 7.6a refers to the displacement of CFM relative to the input end.

In this work, the selected linear actuator, i.e., voice coil motor (VCM), can provide the maximum actuation force of 81 N and maximum displacement of 12.7 mm. The output force property of the VCM is obtained as shown in Fig. 7.6b. Referring to Fig. 7.6b, the actuator can provide an output force greater than 0.7 N ranging from 0 to 12.59 mm, which is sufficient to offer the predicted displacement of 2.5 mm. The stroke requirement of the VCM is verified in the following discussion.

7.3.2 Gripping Force and Gripping Stroke Consideration

The gripping force and gripping stroke are important characteristics of the gripper. They are determined by the mechanical property of the CFM and the gripper jaws. The mechanical property of the designed CFM is predicted by FEA simulation study earlier. The property of the gripper jaws is analyzed as follows.

The amplification ratio of the gripper jaw can be determined by using Eqs. (7.3) and (7.4). According to the parameters given in Table 7.1, the actual values of the parameters for the movement arm of the lever can be calculated as follows.

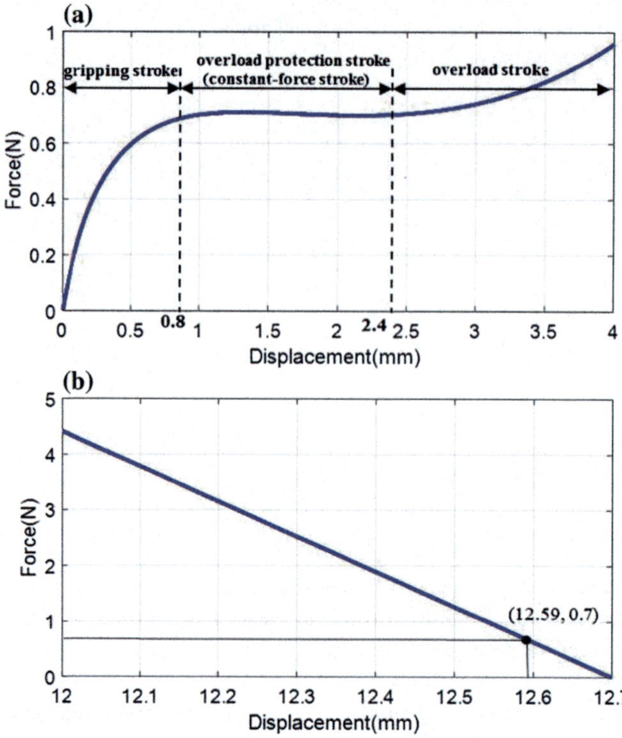

Fig. 7.6 Simulation results of **a** constant-force property of CFM; **b** output force property of the selected VCM

$$d' = d\sin\phi - \frac{w}{2} + \frac{b}{2}\cos\phi = 8.64\ \text{(mm)} \tag{7.6}$$

$$L' = L - \frac{b}{2}\sin\phi + w\cos\phi = 50.12\ \text{(mm)} \tag{7.7}$$

Then, based on Eq. (7.4), the amplification ratio of a gripper jaw is calculated as β = 11.6.

To verify the analytical model result, FEA simulation study is conducted and the result is depicted in Fig. 7.7a. By applying 1 mm displacement on the input end of the gripper jaws, the output displacement of 11.3 mm indicates that the amplification ratio of one jaw is 11.3. Hence, the result of the analytical model agrees well with that of simulation with a small deviation of 2.7%, which verifies the accuracy of the derived model.

After getting the magnification of the gripper jaws, the gripping force can be determined by Eqs. (7.8) and (7.9) below.

$$F_{gripping} \cdot L = F_{input} \cdot d \tag{7.8}$$

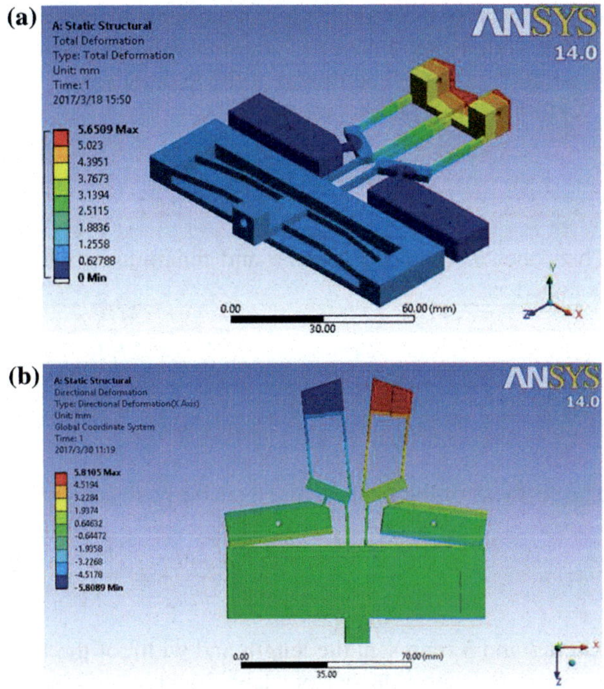

Fig. 7.7 Simulation results of **a** displacement amplification analysis and **b** gripping range analysis

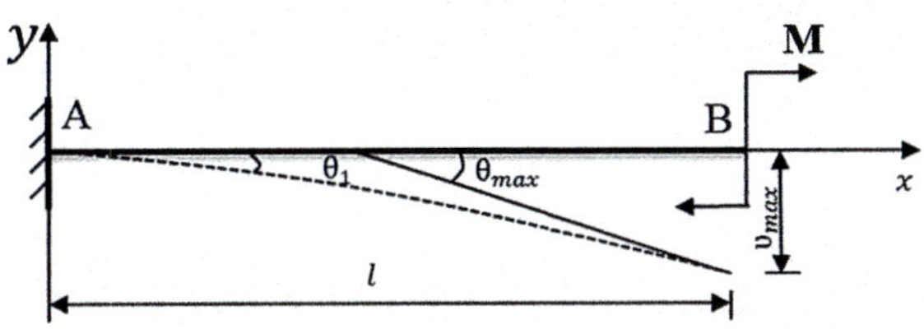

Fig. 7.8 Illustration of deformation for the cantilever beam

$$F_{gripping} = F_{input} \cdot \frac{L}{d} = F_{input} \cdot \beta_0 \tag{7.9}$$

where the input force F_{input} of the jaws is the output force of the CFM.

The gripping stroke is related to the stiffness of the gripper jaws. Through mechanics analysis, it is derived that the stiffness of the jaw is mainly governed by the flexible hinges on the pivot of lever. The flexible hinges can be considered as cantilever beams, and its force analysis is illustrated in Fig. 7.8.

According to the force condition, an approximate differential equation of the beam deflection can be obtained as follows.

$$\frac{d^2v}{dx^2} = \frac{M}{EI} \tag{7.10}$$

where v represents the deflection of the beam in transverse direction (y-axis), E and I denote Young's modulus and rotational moment of inertia of the beam, respectively. In addition, x is the coordinate of the lateral axis.

After the integral operation, yields

$$v = \frac{M}{2EI} \cdot x^2 \tag{7.11}$$

Then, the maximum deflection and maximum rotation angle of the beam can be obtained as:

$$v_{max} = \frac{Ml^2}{2EI} \tag{7.12}$$

$$\theta_{max} = 2 \cdot \theta_1 = \frac{Ml}{EI} \tag{7.13}$$

The stiffness of the flexible hinges is derived as follows.

$$K = \frac{M}{\theta_{max}} = \frac{Ebh^3}{12l} \tag{7.14}$$

where l and b represent the length and width of the flexible hinges, respectively. In addition, h denotes the out-of-plane thickness of hinges.

Considering that the designed gripper jaws contain two identical flexible hinges, the stiffness can be calculated as follows.

$$K_{fingertips} = \frac{Ebh^3}{6l} \tag{7.15}$$

Therefore, the gripping displacement and the driving displacement can be calculated below.

$$d_{fingertips} = \frac{F_{CFM}}{K_{fingertips}} \cdot \beta_0 \tag{7.16}$$

$$d_{driving} = d_{CFM} + \frac{F_{CFM}}{K_{fingertips}} \tag{7.17}$$

To verify the gripper performance, FEA simulation study is conducted and the results are shown in Figs. 7.7b and 7.9. Figure 7.9a depicts the relationship between the force and displacement. For comparison, the relationship between the driving displacement and gripping force is shown by dotted line in Fig. 7.9a, and its close-up view is given in Fig. 7.9b. We can see that the gripping displacement of the actuator is about 1.9 mm and the overload protection stroke of the actuator is about 1.6 mm ranging from 1.9 to 3.5 mm. That is, the actuator should offer at least an effective stroke of 3.5 mm, which can be easily achieved by the selected VCM. Moreover, through the stroke amplifier of gripper jaws, the output force of CFM can be reduced

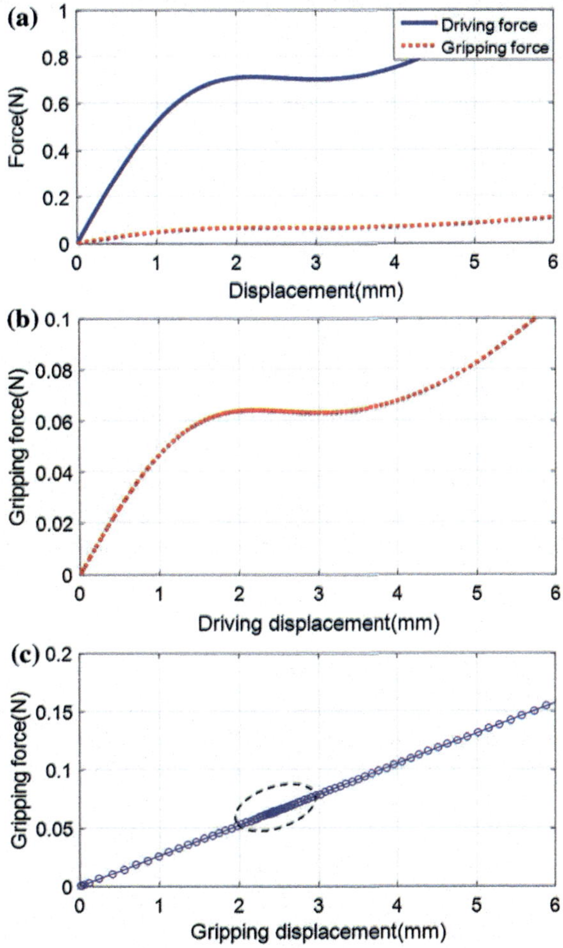

Fig. 7.9 Simulation results of **a** relationship between driving force and driving displacement, **b** relationship between gripping force and driving displacement, and **c** relationship between gripping force and gripping displacement

effectively, whose magnitude is suitable for micro-object manipulation. Figure 7.9c shows the relationship between the gripping force and gripping stroke. At the position around 2.4 mm, the data are very concentrated, which indicates that the gripper jaws are almost stationary. In other words, the gripper lies in overload protection stroke.

The adjusting jaw position is another merit of the presented design. It makes the gripper be suitable for handling objects with various sizes. The adjusting space is called gripping range, which is only related to the parameters of the structure. The extreme position occurs when different parts of the structure just exhibit interference contact during the adjustment, as illustrated in Fig. 7.5b.

The simulation result for the gripping range of the designed gripper is shown in Fig. 7.7b. It is found that the maximum displacement of one jaw is about 5.8 mm. So, the gripping range of the designed gripper can be easily calculated by adding the initial gap between the two jaws to the maximum displacement of the jaws. In the presented case design, the gripping range is from 0 to 13.6 mm corresponding to the original jaw angle adjusted from $-1.65°$ to $4.77°$. It reveals that the developed gripper has a large gripping range and can handle objects with the size ranging from 0 to 13.6 mm.

7.3.3 Parametric Study

The parameters of the gripper are carefully designed to make sure that it offers a desired performance. Nevertheless, the actual values of the parameters are governed by the manufacturing tolerance. Therefore, it is necessary to evaluate the influence of the manufacturing tolerance on the gripper performance.

As compared with other parts, the flexure beams in CFM undergo buckling deformation and are more sensitive to the parameters. Hence, the influences of CFM parameters are mainly discussed here. In particular, the influences of the inclination angle, in-plane width, and out-of-plane thickness are generated by FEA simulation study.

7.3.3.1 Inclination Angle

As the inclination angle of the beam varies from 8.6° to 9.4°, the output force property of the CFM is shown in Fig. 7.10a. The results indicate that, an incremental of 0.2° for the inclination angle leads to an increase of about 0.006 N of the force. Also, the constant-force stroke is increased along with the increasing of the inclination angle. However, the relationships are not of directly proportional.

7.3.3.2 Out-of-Plane Thickness

The out-of-plane thickness of the beam is increased from 2.6 mm to 3.4 mm with an interval of 0.2 mm. The output force property of the CFM is obtained as depicted in Fig. 7.10b. We can observe that the force increases as the thickness increases. The relation between the force and out-of-plane thickness is of approximately directly proportional. In addition, the change of the out-of-plane thickness imposes nearly no effect on the constant-force stroke.

7.3.3.3 In-Plane Width

The in-plane width of the beam is varied from 0.5 to 1.4 mm gradually with a step of 0.1 mm. Figure 7.10c shows the obtained force–deflection relationships. In comparison with the other parameters, the in-plane width induces the largest influence

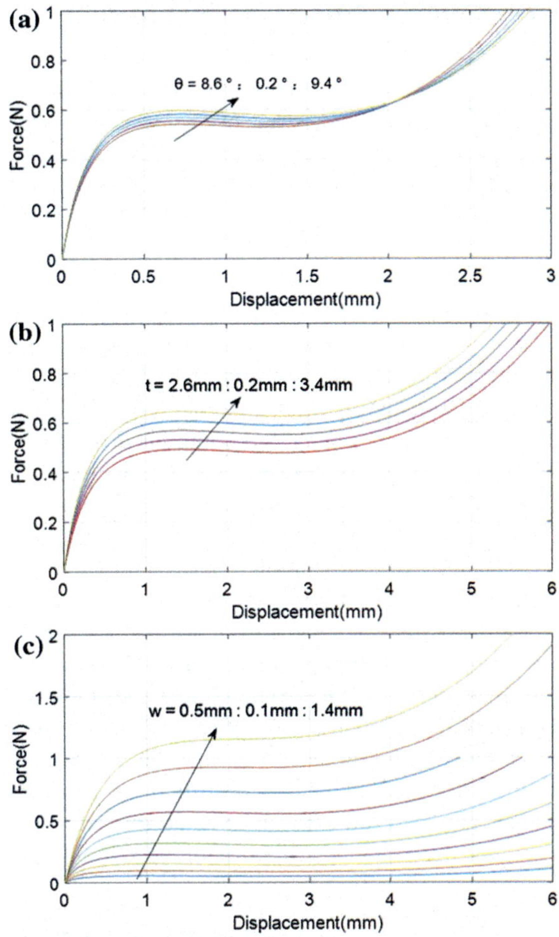

Fig. 7.10 Simulation results of force–displacement relationship when changing **a** inclination angle, **b** out-of-plane thickness, and **c** in-plane width of the flexure beam in CFM

on the force. Specifically, the force increases as the width increases. The larger the beam width, the larger the influence on the force.

From the perspective of machining process, the inclination angle of the bistable beams is difficult to be guaranteed. Fortunately, it has small influence on the performance of the gripper. Moreover, the in-plane width of the beam has the greatest influence on the performance of CFM, while it is easy to be guaranteed by the machining process. The aforementioned analysis provides a guideline of tolerance selection for the manufacturing process.

7.4 Experimental Investigations

In this section, a prototype of the gripper is developed, and its performances are tested by carrying out a series of experimental studies.

7.4.1 Prototype Development

The experimental setup of the developed gripper is shown in Fig. 7.11. The prototype of the gripper is fabricated using the material of ABS plus by a 3D printer (model: uPrint-SE plus, from Stratasys Ltd.). The advantage of 3D printing lies in lower implementation cost and easier to fabricate for performance verification of the gripper. In the experiment, the gripper is driven by a VCM (model: NCC05-18-060-2X, from H2W Technologies, Inc.). The VCM adopts an input current and produces an output force. To transform the force output of VCM into displacement output, a multistage compound parallelogram flexure stage [32] is employed as a guiding mechanism. The combination of VCM and guiding mechanism acts as a driving unit of the gripper. The driving displacement and gripping displacement are measured by two laser displacement sensors (model: LK-H055, from Keyence Corporation), which provide a resolution of 25 nm within a measurement range of 20 mm. A force sensor (model: LSB200, from Futek Inc.) is mounted between the actuator and driving end of the gripper to measure the actuation force of stage. In addition, the force sensor can also be mounted between the gripper jaws to measure the gripping force for the gripper. A real-time controller (model: CRIO-9075, from National Instruments Corp.) is used to produce excitation signals and acquire the sensor readings. A microscope is adopted to observe the gripping process.

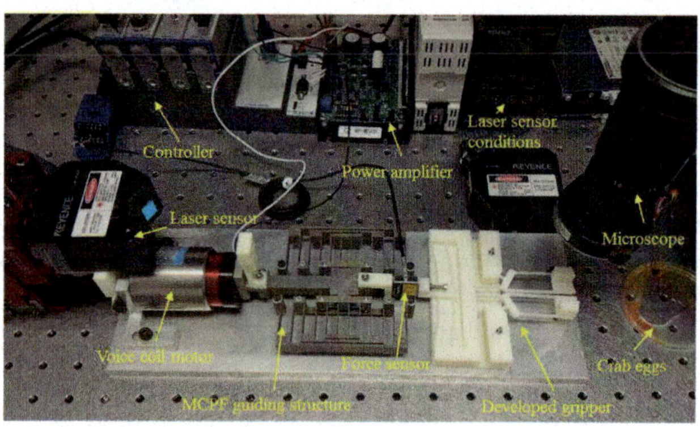

Fig. 7.11 Experimental setup of the constant-force gripper

7.4.2 *Performance Testing Results*

7.4.2.1 Performance of the CFM

First, the performance of the CFM is examined by carrying out experimental study. A 0.1-Hz quasi-static sinusoidal signal is applied to the VCM actuator. By driving the VCM with a voltage amplitude of ± 10 V, a motion range of 5.61 mm is achieved for the driving unit, as shown in Fig. 7.12a. The experimental results of ten periods are gathered to eliminate the contingency and to show the real performance of the designed gripper. The experimental results of the driving force are shown in Fig. 7.12b, where the model predicted result is used for comparison. The constant-force value is about 0.72 N in a motion range of 2–4.2 mm, which is 1.4% larger than the predicted result of 0.71 N. During the constant-force stroke, the force fluctuation is about 0.03 N (i.e., 4.2% of the output force value), which shows a good constant output force property.

As for the CFM, the experimental result of the constant-force stroke is about 1.8 mm, ranging from 2.2 to 4.0 mm. The experimental result is 12.5% larger than the predicting value of 1.6 mm. The starting and ending positions of constant force are a little different between the experimental and model predicted results. The difference is small, which arises from the modeling simplification and machining error of the gripper.

The movement results of CFM are shown in Fig. 7.12c, where the lateral axis represents the driving displacement. Before entering the constant force process, the displacement of CFM increases gradually and the maximum value is about 0.16 mm, which is 14.3% larger than the predicted result of 0.14 mm. In theory, the CFM will keep still in constant stroke, but it exhibits some reverse movement in experiment. Even so, the reverse displacement is below 0.02 mm, which is relatively small and can be approximately considered as zero. This issue is caused by the machining error, which generates a not absolutely constant force.

In addition, in ten periods of motion tests, the displacements of grasp-and-release movements are not totally overlapped. The displacement of grasp movement is a little larger than that of release movement, especially in the starting position of the constant-force stroke. Material relaxation/creep may also cause the discrepancy in the model prediction and experimental measurement in the repeated test. It is notable that the positioning accuracy and repeatability of the gripper can be improved by using a closed-loop position control.

7.4.2.2 Performance of the Gripper

The constant-force capacity of the gripper has also been tested. The experimental results of the gripping stroke and gripping force testing are shown in Fig. 7.13. The lateral axis represents the driving displacement. It is obvious that the gripping displacement and gripping force increase gradually before the driving displacement

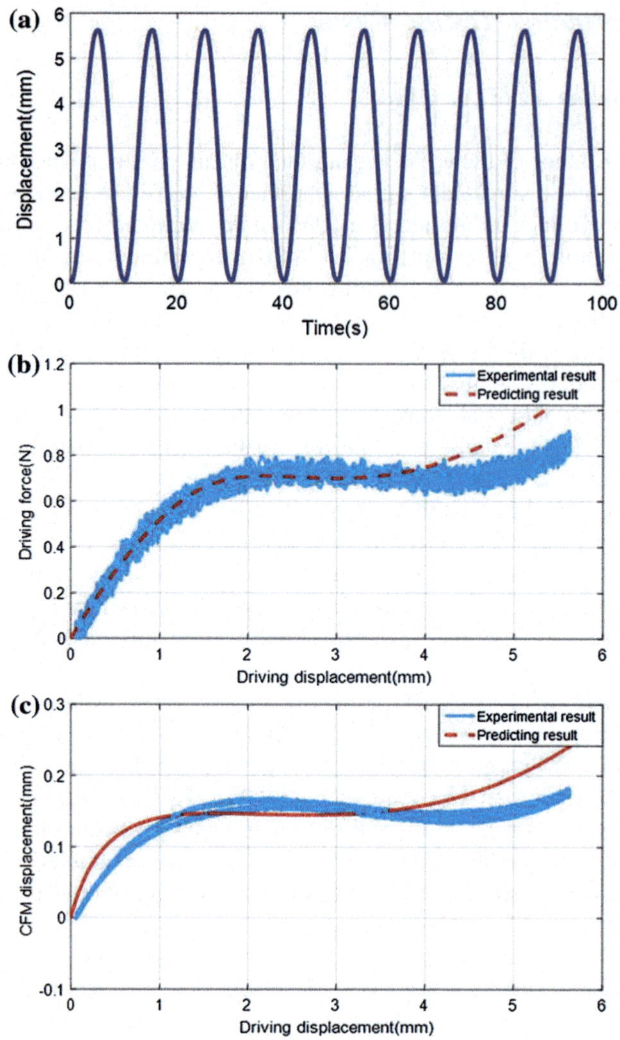

Fig. 7.12 Experimental results of **a** input displacement, **b** driving force, and **c** displacement of the CFM

arrives at the position of 2.1 mm. After that, the gripping force also keeps a constant until the driving displacement arrives at 3.5 mm. Each jaw of the gripper has a 0.81-mm motion range. So, the two-jaw gripper has a gripping stroke of 1.62 mm, which is 26.3% smaller than the model predicted value of 2.2 mm. The constant gripping force value is about 0.071 N, which is 16.4% larger than the model result of 0.061 N. The experimental and predicted results are consistent, which verifies the correctness of the models.

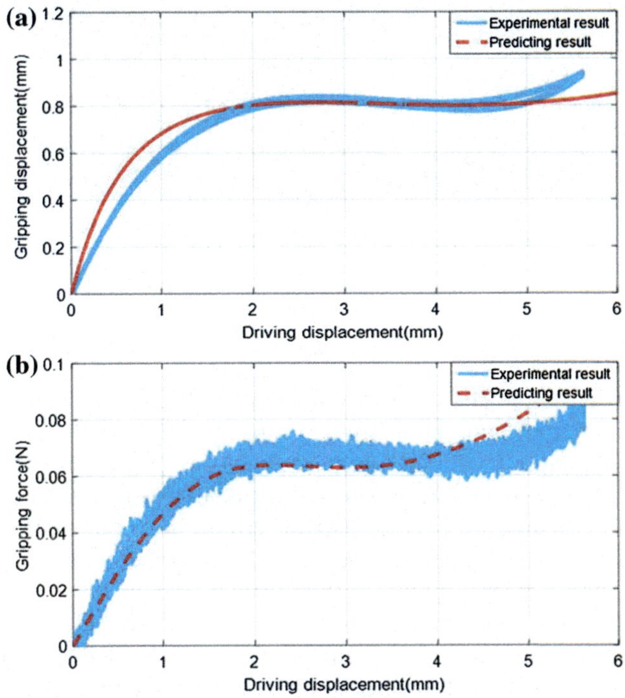

Fig. 7.13 Experimental results of **a** gripping displacement and **b** gripping force of the constant-force gripper

In the constant-force stroke, the gripping force will not change as the increasing of the driving displacement. For this reason, it can be used for overload protection in practical application. During the ten times grasp-and-open movement, the value of the maximum gripping force floats around 0.071 N with an error less than ±0.01 N, which indicates a stable force property of the gripper. The gripping displacement also suffers from misalignment of back-and-forth movement of the driving unit. It mainly occurs after 4.4 mm of driving displacement, which is out of working range and can be ignored. Thus, the gripping movement reveals a good repeatability.

Thus, the experimental results show that designed gripper demonstrates a good constant-force performance. The experimental results also verify the accuracy of the development analytical model.

7.4.3 Biological Gripping Application

The designed gripper can be used for micromanipulation tasks. For demonstration, the gripper is employed for biological cell microinjection. Crab eggs are adopted

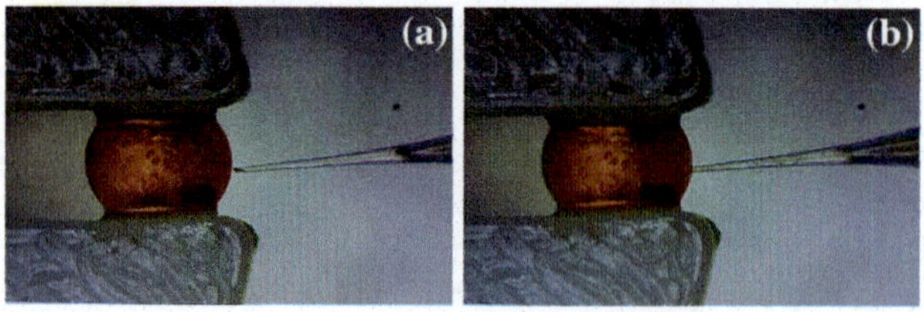

Fig. 7.14 Snapshots of **a** before and **b** after microinjection process of crab egg

as biological cells and held by the gripper. The snapshots of the cell microinjection process are shown in Fig. 7.14.

It is found that the cell does not move or break in the experiment. This indicates that the gripping force is appropriate for the crab cell microinjection. If the gripping force is too small, the cell will move during the injection. Otherwise, if the gripping force is too large, the cell will break before injection or explore during the injection.

It is noted that the two jaws in the designed gripper do not move in pure parallel direction, which can induce a slipping force on the objects and push the objects out of the gripper jaws in theory. While the experimental results show that by carefully designing the gripping force of the gripper, the friction between the gripper jaws and object can overcome the influence of slipping force. Hence, the slipping force has nearly no influence on the performance of the gripper during the cell injection.

7.4.4 Comparison Study Result

As discussed in introduction, the developed gripper not only generates a constant gripping force, but also reduces the gripping force to a very small magnitude with a compact size. In this section, a comparison study is conducted to show the advantages of the developed gripper.

As shown in the literature [15], it is challenging to achieve a low magnitude of the constant force for micromanipulation application. In the literature, the grippers as reported in [15] (gripper A) and [16] (gripper B) can generate the smallest gripping force about 0.7 N. By contrast, the gripper presented in this chapter can produce a gripping force of 0.071 N, which has been significantly reduced by ten times versus existing works. To have an intuitive comparison of the force magnitude, a cell gripping experiment is conducted using gripper A [15] and the proposed one.

The experimental results are shown in Fig. 7.15 by adopting crab egg embryos as micro-objects. The original states of the crab eggs are shown in Fig. 7.15a, c, respectively. The photographs after gripping by the two grippers are shown in Fig. 7.15b, d, respectively. It is observed that the cell deformation as shown in Fig. 7.15d is much smaller than that in Fig. 7.15b. The deformation rates as shown in Fig. 7.15b, d are about 77% and 18%, respectively. In cell micromanipulation, large deformation will break the equilibrium state inside the cell. In the experiment, the property of

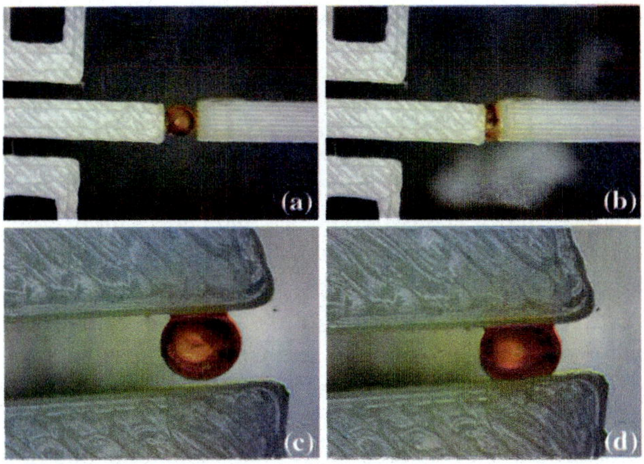

Fig. 7.15 Cell gripping test results by using existing gripper **a–b** and proposed gripper **c–d**. **a**, **c**: Original state of crab egg; **b**, **d**: cell deformation after gripping

crab egg in Fig. 7.15b has been changed and it cannot recover to its original shape after release. On the contrary, the crab egg in Fig. 7.15d can gradually recover to its original shape after release. This reveals the significance of reducing the gripping force and exhibits the advantage of the developed constant-force gripper over the state-of-the-art design.

7.4.5 Further Discussion

The size of the gripper is another important performance criterion. The grippers A and B can also reduce their gripping force into the same level as the designed gripper. But their structures have to be increased to a very large size. In addition, reference [27] reports a promising constant-force compliant gripper (gripper C), which can handle objects with various sizes. For demonstration, the performance of the developed gripper is compared with those of the existing grippers A, B, and C, as tabulated in Table 7.2.

Table 7.2 Main performances of different constant-force grippers

Gripper	Size (mm^2)	Force (N)	Stroke (mm)	Compactness ($N{\cdot}mm^2$)
Gripper A [15]	2210 × 1114	0.75	2.2	1846455
Gripper B [16]	79 × 68	0.55	0.40	2954.6
Gripper C [27]	140 × 60	23.5	–	197400
This work	100 × 109	0.071	2.2	773.9

For a quantitative comparison, a compactness index (γ) is introduced to indicate the compactness of the gripper as follows.

$$\gamma = F_{gripping} \cdot S \tag{7.18}$$

where $F_{gripping}$ represents the value of its constant gripping force and S represents the planar area of the gripper. A smaller index γ indicates a more compact structure of the gripper for micro-object manipulation.

The γ values of grippers A, B, and C are calculated as 1846455, 2954.6, and 197400 N·mm^2, respectively. The index γ of the designed gripper is 773.9 N·mm^2. Therefore, as compared to grippers A, B, and C, the compactness of the designed gripper has been enhanced by 2385, 3.8, and 255 times, respectively. It indicates that in comparison with existing designs, the proposed constant-force gripper can achieve the same force performance with much lower cost on material and fabrication.

7.5 Conclusion

This chapter presents the design, analysis, and experimental study of a flexure-based robotic gripper with a constant gripping force, large handling size, and compact physical dimension. Analytical model of the gripper has been developed. Model results show that the driving force, driving stroke, gripping force, gripping stroke, and gripping range are reasonable. Experimental results show that the fabricated prototype delivers a constant gripping force of 0.071 N in the gripping displacement of 1.62 mm along with a small force fluctuation. The gripper has a large gripping range and can handle objects with the size ranging from 0 to 13.6 mm. In addition, the developed gripper exhibits a more compact dimension in comparison with existing constant-force grippers. Furthermore, a comparison experimental study has been conducted to reveal the advantages of the developed constant-force gripper for micromanipulation tasks. The results have verified the advantages of the small magnitude of gripping force and compact structure size.

References

1. Beyeler, F., Neild, A., Oberti, S., Bell, D.J., Sun, Y., Dual, J., Nelson, B.J.: Monolithically fabricated microgripper with integrated force sensor for manipulating microobjects and biological cells aligned in an ultrasonic field. J. Microelectromechanical Syst. **16**(1), 7–15 (2007)
2. Chen, C.C., Lan, C.C.: An accurate force regulation mechanism for handling fragile objects using pneumatic grippers. In: Proc of 2016 IEEE International Conference on Advanced Intelligent Mechatronics (AIM), pp. 389–394. Banff, Alberta, Canada (2016)
3. Chen, G., Ma, Y., Li, J.: A tensural displacement amplifier employing elliptic-arc flexure hinges. Sensors. Actuators A: Phys. **247**, 307–315 (2016)

4. Chen, T., Wang, Y., Yang, Z., Liu, H., Liu, J., Sun, L.: A PZT actuated triple-finger gripper for multi-target micromanipulation. Micromachines **8**(2), 33 (2017)
5. Choi, S., Han, S., Lee, Y.: Fine motion control of a moving stage using a piezoactuator associated with a displacement amplifier. Smart Mater. Struct. **14**(1), 222 (2004)
6. Hao, G., Mullins, J., Cronin, K.: Simplified modelling and development of a bi-directionally adjustable constant-force compliant gripper. Proc. Inst. Mechanical Eng. Part C: J. Mechanical Eng. Sci. **231**(11), 2110–2123 (2017)
7. Holst, G.L., Teichert, G.H., Jensen, B.D.: Modeling and experiments of buckling modes and deflection of fixed-guided beams in compliant mechanisms. J. Mechanical Des. **133**(5), 051,002 (2011)
8. Huang, H., Sun, D., Su, H., Mills, J.K.: Force sensing and control in robot-assisted suspended cell injection system. In: Advances in Robotics and Virtual Reality, pp. 61–88. Springer (2012)
9. Huang, L.Y., Lin, Y.C., Liu, Y.C., Su, J.Y., Lin, P.C.: A manipulator with a depth sensor and an underactuated and tactile gripper for identifying and grasping objects of various shapes and sizes. In: Proc. of The 14th IFToMM World Congress, pp. 1–8 (2015)
10. Kim, H., Kim, J., Ahn, D., Gweon, D.: Development of a nanoprecision 3-DOF vertical positioning system with a flexure hinge. IEEE Trans. Nanotechnology **12**(2), 234–245 (2013)
11. Kim, K., Liu, X., Zhang, Y., Sun, Y.: Nanonewton force-controlled manipulation of biological cells using a monolithic mems microgripper with two-axis force feedback. J. Micromechanics Microengineering **18**(5), 055,013 (2008)
12. Lamers, A.J., Sanchez, J.A.G., Herder, J.L.: Design of a statically balanced fully compliant grasper. Mechanism Machine Theory **92**, 230–239 (2015)
13. Liu, Y., Xu, Q.: Design and analysis of a micro-gripper with constant force mechanism. In: Proc. of 2016 12th World Congress on Intelligent Control and Automation (WCICA), pp. 2142–2147 (2016)
14. Liu, Y., Xu, Q.: Design of a compliant constant force gripper mechanism based on buckled fixed-guided beam. In: Proc. of International Conference on Manipulation, Automation and Robotics at Small Scales (MARSS), pp. 1–6 (2016)
15. Liu, Y., Xu, Q.: Design of a 3D-printed polymeric compliant constant-force buffering gripping mechanism. In: Proc. of 2017 IEEE International Conference on Robotics and Automation (ICRA), pp. 1–6 (2017)
16. Liu, Y., Zhang, Y., Xu, Q.: Design and control of a novel compliant constant-force gripper based on buckled fixed-guided beams. IEEE/ASME Trans. Mechatronics **22**(1), 476–486 (2017)
17. Lobontiu, N., Cullin, M., Petersen, T., Alcazar, J.A., Noveanu, S.: Planar compliances of symmetric notch flexure hinges: the right circularly corner-filleted parabolic design. IEEE Trans. Autom. Sci. Eng. **11**(1), 169–176 (2014)
18. Ma, C.Z., Du, J.S., Liu, Y.Y., Chu, Y.K.: Overview of micro-force sensing methods. Appl. Mech. Mater. **462**, 25–31 (2014)
19. Medina, L., Gilat, R., Ilic, B.R., Krylov, S.: Open-loop, self-excitation in a bistable micromechanical beam actuated by a dc electrostatic load. In: Proc. of 2017 IEEE 30th International Conference on Micro Electro Mechanical Systems (MEMS), pp. 785–788 (2017)
20. Midha, A., Bapat, S.G., Mavanthoor, A., Chinta, V.: Analysis of a fixed-guided compliant beam with an inflection point using the pseudo-rigid-body model concept. Journal of Mechanisms and Robotics **7**(3), 031,007 (2015)
21. Mølhave, K., Hansen, O.: Electro-thermally actuated microgrippers with integrated force-feedback. J. Micromechanics Microengineering **15**(6), 1265 (2005)
22. Pham, H.T., Wang, D.A.: A constant-force bistable mechanism for force regulation and overload protection. Mechanism Machine Theory **46**(7), 899–909 (2011)
23. Reddy, A.N., Maheshwari, N., Sahu, D.K., Ananthasuresh, G.K.: Miniature compliant grippers with vision-based force sensing. IEEE Trans. Rob. **26**(5), 867–877 (2010)
24. Stavenuiter, R.A.J., Birglen, L., Herder, J.L.: A planar underactuated grasper with adjustable compliance. Mechanism Machine Theory **112**, 295–306 (2017)
25. Tolman, K.A., Merriam, E.G., Howell, L.L.: Compliant constant-force linear-motion mechanism. Mechanism Machine Theory **106**, 68–79 (2016)

26. Tolou, N., Pluimers, P., Jensen, B.D., Magleby, S., Howell, L.L., Herder, J.L.: Near-zero-stiffness linear motion stage with high orthogonal and out-of-plane stiffness. Proc of 1st DSPE Conf. on Precision Mechatronics, pp. 1–2. Deurne, The Netherlands (2012)
27. Wang, J.Y., Lan, C.C.: A constant-force compliant gripper for handling objects of various sizes. J. Mechanical Des. **136**(7), 071,008 (2014)
28. Wang, P., Xu, Q.: Design of a flexure-based constant-force XY precision positioning stage. Mechanism Machine Theory **108**, 1–13 (2017)
29. Wei, Y., Xu, Q.: An overview of micro-force sensing techniques. Sensors Actuators A: Phys. **234**, 359–374 (2015)
30. Wu, Q., Wang, X., Chen, B., Wu, H., Shao, Z.: Development and hybrid force/position control of a compliant rescue manipulator. Mechatronics **46**, 143–153 (2017)
31. Xie, Y., Sun, D., Tse, H.Y.G., Liu, C., Cheng, S.H.: Force sensing and manipulation strategy in robot-assisted microinjection on zebrafish embryos. IEEE/ASME Trans. Mechatronics **16**(6), 1002–1010 (2011)
32. Xu, Q.: Design, testing and precision control of a novel long-stroke flexure micropositioning system. Mechanism Machine Theory **70**, 209–224 (2013)
33. Xu, Q.: Design of a large-stroke bistable mechanism for the application in constant-force micropositioning stage. J. Mechanisms and Robotics **9**(1), 011,006 (2017)
34. Zhang, D., Zhang, Z., Gao, Q., Xu, D., Liu, S.: Development of a monolithic compliant SPCA-driven micro-gripper. Mechatronics **25**, 37–43 (2015)

Chapter 8
Design and Implementation of a Force-Sensing MEMS Microgripper

Abstract This chapter presents the design, simulation, fabrication, and testing processes of a microelectromechanical systems (MEMS) microgripper, which integrates an electrostatic actuator and a capacitive force sensor. One advantage of the presented gripper is that the gripping force and interaction force in two orthogonal directions can be respectively detected by a single force sensor. The gripper exhibits a simple structure and compact footprint. The MEMS gripper is fabricated by SOI-MUMPs process. The performance of the designed gripper is verified by conducting finite element analysis (FEA) simulation and experimental studies. Moreover, the demonstration of bio-cellulose gripping confirms the feasibility of the developed gripper device.

8.1 Introduction

Microelectromechanical systems (MEMS) microgripper plays an important role in micromanipulation and biomedical areas. It is adopted to implement the grasp of tiny objects. Typical applications include delivering micro-objects [4], assembling microelements [6], and characterizing bio-materials [2]. The mechanical structure of a microgripper is usually designed as a compliant mechanism [9], which works based on elastic deformation of the material. This chapter introduces the design and fabrication of a MEMS microgripper which possesses an electrostatic actuator and a capacitive force sensor.

A number of actuation approaches have been adopted to drive MEMS microgrippers, e.g., electrothermal actuation [24], piezoelectric actuation [20], shape memory alloy actuation [1], and electrostatic actuation [15]. Among these driving methods, electrothermal actuators have the ability of providing a large force with a pretty low input voltage, usually below 10 V. Whereas the relatively high work temperature (over about 400 K) becomes a big issue for handling materials which are sensitive to temperature. Piezoelectric actuators have the advantages of fast response speed and good bandwidth. But the nonlinear hysteresis of the displacement response complicates the gripper's control design process. Regarding the shape memory alloy actuator, it exhibits some merits like large stroke, high force density, and large recovery force.

Q. Xu, *Micromachines for Biological Micromanipulation*,
https://doi.org/10.1007/978-3-319-74621-0_8

Yet, the hysteresis effect and the high power consumption restrict its application in some scenarios. In contrast, electrostatic actuators offer the merits of no hysteresis, simple structure, and fast response speed, and it is able to generate a sufficient displacement for grasping operation. Thus, the electrostatic actuator is chosen for driving the gripper in this work.

In order to prevent the grasped microsamples from damaging, MEMS microgrippers can be equipped with force sensors for detecting the force between the grasped objects and gripper tips. In the literature, piezoelectric sensor [11], electrothermal sensor [22], pizeoresistive sensor [14], and capacitive sensor [10] are commonly used in MEMS microgripper design. Generally, piezoelectric sensor provides a wide bandwidth, while the limitation comes from its large dimension and complicated fabrication process. Concerning electrothermal sensor, it supplies a high resolution and owns a compact size. But it also has some drawbacks like the relatively high energy consumption and temperature sensitivity. Piezeoresistive sensor introduces flexibility and high bandwidth for the microgipper. Yet, its problem mainly comes from the hysteresis effect. Alternatively, a capacitive force sensor is designed in this work, as it offers some advantages such as simple structure, fast response speed, large bandwidth, and high sensitivity. Thus, capacitive force sensor is designed for the microgripper in this work.

In addition, to guarantee the reliability and manipulation accuracy of a MEMS microgripper, it is necessary to measure both gripping force and interaction force in orthogonal directions [21]. The gripping force is the force generated between the gripper tips and grasped objects. Majority of the existing force-sensing microgrippers have the capability of detecting the force in gripping direction by using a force sensor [5, 18]. The interaction force is caused between the gripper tips and the environment. It is necessary to measure the interaction force for determining the contact moment in real time in practical grasp operation [7]. However, existing microgrippers rarely have the function of interaction force sensing [12, 19]. Moreover, the previous work complicates the mechanical structure design because they use two separate force sensors. In this chapter, a MEMS microgripper is reported which possesses an electrostatic actuator and a capacitive force sensor. The single force sensor is designed to have the capability of dual-axis force sensing, which can detect the forces of gripping and interaction in two orthogonal directions in sequence.

8.2 Mechanism Design of the Microgripper

The mechanical design of the MEMS microgripper is illustrated in Fig. 8.1. The gripper structure consists of two parts, i.e., left and right parts, which provide the functions of actuating and sensing, respectively [23]. In the left actuating part, electrostatic comb drive is used. It is connected to the left arm by four leaf flexures for executing the grasp operation. The left arm of the gripper is designed based on the lever amplification principle, and it is linked to the movable shuttle of the comb drive by two folded flexures. The use of folded flexure contributes to the generation of a

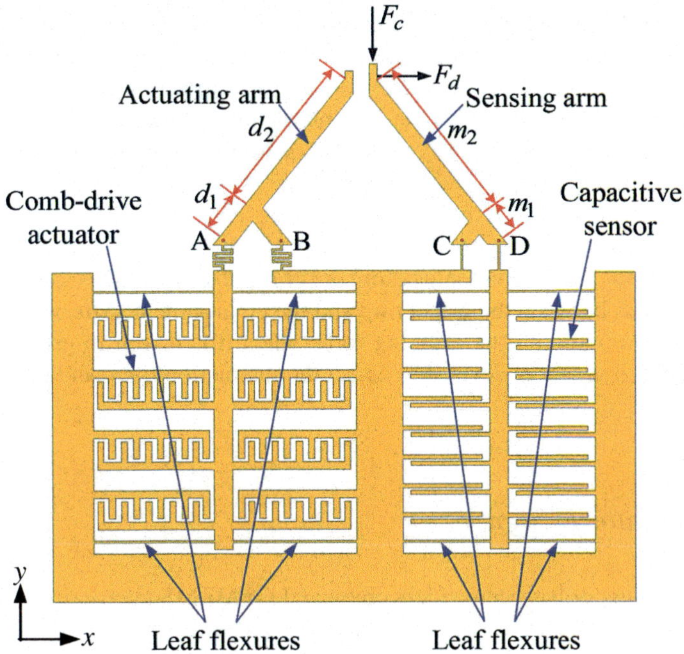

Fig. 8.1 Schematic diagram of the MEMS microgripper

larger deformation under the stress constraint of the material. When the deformation is transmitted from comb drive to the left bottom arm (point A), the output displacement at the left tip can be proportionally amplified. In the meantime, the driving force F_{in} is delivered from the electrostatic comb drive to the left tip. Thus, the driving force is transformed as the gripping force F_d. It is notable that the left gripper tip does not experience a pure translational grasping movement. While comparing with the gripping range of hundreds of micrometers, the slightly bias motion (in the direction orthogonal to gripping direction) of left tip does not influence the gripping operation [3].

Concerning the right sensing part, the structure of the right arm is similar to that of the left one. The right sensing arm is connected to a capacitive sensor, which is connected to the fixing base by four leaf flexures. Based on the same motion guiding principle, the supporting flexures in actuation and sensing parts provide a liner guide for the actuator and sensor, respectively. Moreover, the displacement of the right tip in x-axis will be transferred as the y-axis displacement of sensor plates by the right arm. Hence, the gripping force F_d is measured by the capacitive sensor.

It is seen that the right sensing tip is slightly longer than the left tip. When the gripper is translated to approach the environment in y-axis direction, the contact will be established between the right tip and the environment first. The contact induces the interaction force F_c on the right tip, which causes the rotation of the right tip around the point C as shown in Fig. 8.1. This rotation will cause the translation of point D along y-axis. As a result, the interaction force produces a translation of the sensor moving part along y-axis direction. In this way, the interaction force F_c can also be detected by the capacitive sensor. When the interaction force is not exactly orthogonal to the gripping force, a component of the interaction force will be measured by the sensor.

Therefore, the gripper has the capability of detecting the gripping and interaction forces by using a single force sensor. In this work, the force sensor is designed to detect the environmental interaction force before the grasp operation, so as to determine the relative positions between the gripper tip and the environment. Afterward, the force sensor is used to measure the gripping force when it is driven to grasp an object. That is, the force sensor is devised to detect the interaction force and grasp force in sequence.

8.2.1 *Actuator Design*

Electrostatic actuator has been widely employed for MEMS microgripper actuation. According to the working schemes, this actuator can be classified into two types, i.e., transverse comb drive and lateral comb drive. To generate a larger stroke, the lateral comb drive is selected in this work. The working principle of the lateral comb drive electrostatic actuator is shown in Fig. 8.2.

8.2.1.1 Actuation Force and Displacement Design

The driving force of the comb drive is determined by:

$$F_{in} = N\varepsilon t \frac{V^2}{p} \tag{8.1}$$

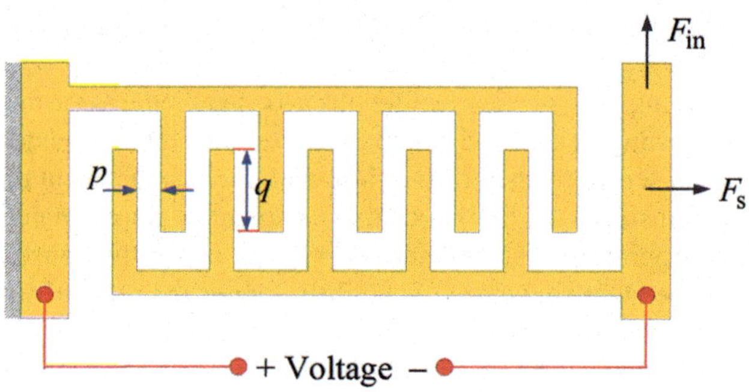

Fig. 8.2 Fundamental principle of lateral type of electrostatic comb drive actuator

where $\varepsilon = 8.85\times10^{-12}$ $C^2/(Nm^2)$ is the permittivity of air, t is the out-of-plane thickness of the gripper, p represents the gap distance between two adjacent plates, V denotes the input voltage, and N is the number of comb teeth pairs.

Referring to Fig. 8.1, the stiffness of the four actuation flexures can be calculated as follows.

$$K_l = 4E\frac{w_a^3 t}{L_a^3} \tag{8.2}$$

where E denotes the Young's modulus of silicon, L_a and w_a represent the length and width of the leaf flexures, respectively. In addition, t is the out-of-plane thickness for the gripper mechanism.

Driving by a voltage, the driving displacement can be derived below.

$$D_{in} = \frac{F_{in}}{K_l} \tag{8.3}$$

The lever amplification ratio of the left actuating arm is approximately governed by

$$R_a = \frac{d_2}{d_1} \tag{8.4}$$

where d_1 and d_2 represent the lengths of the shorter and longer arm of left lever, respectively. The angle between the point going from d_1 to d_2 and point B is around 90°. Simulation study reveals that whether the angle is exactly 90° or not has slight influence on the amplification ratio (8.4).

The displacement at the left tip, which is also called gripping range, can be calculated by

$$D_{out} = R_a D_{in} \tag{8.5}$$

Additionally, in view of Eqs. (8.1) and (8.4), the gripping force at the left tip can be computed by

$$F_d = \frac{F_{in}}{R_a} \tag{8.6}$$

It is notable that Eq. (8.6) is an approximate expression of the transmitted force, as the deformations of the flexures at points A and B cause a slight reduction of the actual force.

Taking into account Eqs. (8.3), (8.5), and (8.6), the stiffness at the left tip can be derived as follows.

$$K_a = \frac{F_d}{D_{out}} = \frac{F_{in}}{D_{in}R_a^2} = \frac{K_l}{R_a^2} \tag{8.7}$$

8.2.1.2 Side Instability Effect

Comb drive actuator not only creates the driving force F_{in} in vertical direction, but also produces an electrostatic force F_s in horizontal direction. The two forces are denoted in Fig. 8.2. Owing to the electrostatic force, the moving teeth are pushed close to the fixed teeth of comb drive. Hence, the side instability phenomenon needs to be considered during the operation of comb drive actuator.

First, the electrostatic force is expressed as [8, 13]:

$$F_s = \frac{1}{2} N\varepsilon t (D_{in} + q) V^2 \left[\frac{1}{(p-x)^2} - \frac{1}{(p+x)^2} \right] \tag{8.8}$$

where x is the displacement of the movable teeth in transverse direction, q is the overlap length of comb teeth driver in rest condition, and p is the clearance size on either side as denoted in Fig. 8.2.

Typically, when the movable teeth are translated to the middle of the gap, the equivalent negative stiffness can be derived as

$$K_s = \left. \frac{\partial F_s}{\partial x} \right|_{x=0} = 2V^2 \frac{N\varepsilon t (D_{in} + q)}{p^3} \tag{8.9}$$

In view of Eq. (8.9), it is observed that K_s will experience an increase as the increasing of the input displacement D_{in}. As a result, K_s should be smaller than the transverse stiffness K_t. In the case of $K_s = K_t$, the side instability effect will be caused. This critical condition can be expressed as follows.

$$2E \frac{w_a t}{L_a} = N\varepsilon t (D_{in} + q) V^2 \frac{1}{p^3} \tag{8.10}$$

To avoid the side instability phenomenon, it is necessary to determine the maximum allowable input voltage of the comb drive. Substituting Eq. (8.3) into Eq. (8.10), we can obtain that

$$V_{max}^2 = \frac{p^2 K_l}{2N\varepsilon t} \left(\sqrt{\frac{2K_t}{K_L} + \frac{q^2}{p^2}} - \frac{q}{p} \right) \tag{8.11}$$

It is notable that the side instability of the electrostatic actuator under more complex loads for the microgripper deserves a further study.

8.2.2 Sensor Design

The working principle of the capacitive force sensor is illustrated in Fig. 8.3.

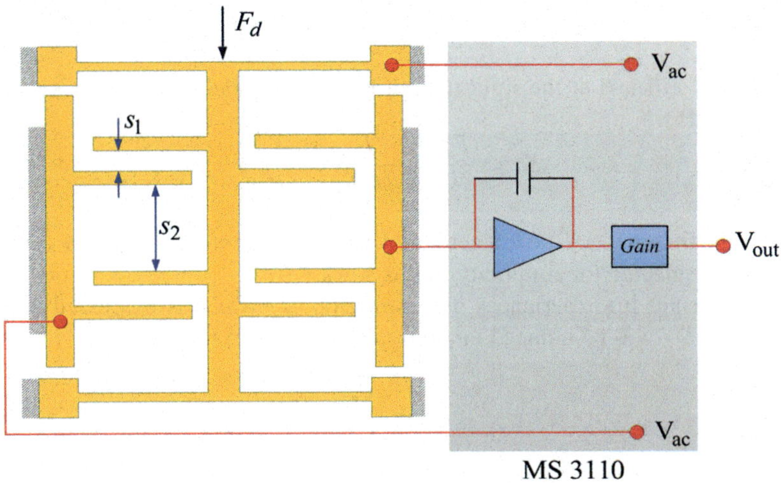

Fig. 8.3 Schematic diagram of capacitive sensor with signal conditioning circuit

8.2.2.1 Gipping Force Sensor Design

When an object is grasped by the two tips of the gripper, the gripping force will be generated and transmitted from the left actuating tip to the right sensing movable plates of the sensor.

The total stiffness of the four leaf flexures in the right part of the gripper can be derived as follows.

$$K_c = 4E\frac{w_s^3 t}{L_s^3} \tag{8.12}$$

where L_s and w_s are the length and width of the sensing leaf flexures, respectively, and t is the out-of-plane thickness of the gripper device.

The amplification ratio of the lever mechanism for the right sensing arm can be approximately calculated by

$$R_s = \frac{m_2}{m_1} \tag{8.13}$$

where m_1 and m_2 are the lengths of the shorter and longer arms of the right lever mechanism, respectively.

Referring to Fig. 8.1, the induced displacement at the right tip D_g is decreased at point D. Specifically, the displacement at point D can be derived as

$$D_f = \frac{D_g}{R_s}. \tag{8.14}$$

In addition, the gripping force F_d at the right tip is transmitted to the force sensor. The sensing force F_t at point D can be computed by

$$F_t = R_s F_d. \tag{8.15}$$

Hence, the stiffness at the right tip can be obtained below by referring to Eqs. (8.14) and (8.15).

$$K_r = \frac{F_d}{D_g} = \frac{F_t}{D_f R_s^2} = \frac{K_c}{R_s^2} \tag{8.16}$$

Similar to the treatment of the left actuation arm, the deflections of flexures at points C and D are neglected for simplicity in this work.

When the right tip experiences a deformation D_g which is expressed below, a displacement D_f will be induced in the movable plates of the sensor.

$$D_g = \frac{F_d}{K_r} \tag{8.17}$$

By assigning the maximum deformation D_g^{max}, the maximum gripping force can be derived as follows.

$$F_d^{max} = \frac{K_c D_g^{max}}{R_s^2}. \tag{8.18}$$

8.2.2.2 Interaction Force Sensor Design

When the gripper is positioned by a micropositioning stage before the grasp manipulation, an initial contact with the environment can be made, which is useful to determine the relative displacements. When the sensing tip contacts with the environment, the interaction force F_c exerted at the right tip will be transmitted to the point D of the sensor in vertical direction, as shown in Fig. 8.1.

As a result, the movable plates will experience a deformation D_c. Thus, by assigning the maximum displacement D_c^{max}, the maximum interaction force can be approximately determined as follows.

$$F_c^{max} = K_c D_c^{max} \tag{8.19}$$

It is notable that a more detailed analysis is needed to predict the force value in the future work.

8.2.2.3 Output Signal of the Capacitive Sensor

To measure the force exerted by the force sensor, a capacitive-to-voltage converter chip (model: MS3110, from MicroSensors, Inc.) can be adopted. It produces an output signal V_{out} of the capacitive sensor due to the capacitance change:

$$V_{out} \propto G \times \frac{C_1 - C_2}{C_f} \tag{8.20}$$

with

$$C_1 = \varepsilon(\frac{A}{S_1 - \lambda} + \frac{A}{S_2 + \lambda}),\ C_2 = \varepsilon(\frac{A}{S_1 + \lambda} + \frac{A}{S_2 - \lambda}) \tag{8.21}$$

where G and C_f are gain values which can be adjusted by referring to some specific situation. A is the overlapping area of the plates. S_1 and S_2 are two gap distances of the parallel capacitive plates. The output voltage is proportional to the deformation λ [12].

8.3 Performance Estimation with FEA Simulation

Finite element analysis (FEA) is carried out with ANSYS software package to verify the feasibility and performance of the microgripper mechanism. In order to simplify and speed up the simulation process, the electrostatic actuation teeth and capacitive sensing plates are removed. This simplification does not influence the result of static FEA simulation.

In this case design, the main parameters of the gripper are listed in the Table 8.1. Both static and dynamic analyses are conducted to examine the performance of the microgripper in this section.

Table 8.1 Main parameters of the microgripper

Parameter	Description	Value (μm)
L_a	Length of the actuating flexure	1200
L_s	Length of the sensing flexure	800
w_a	Width of the actuating flexure	10
w_s	Width of the sensing flexure	10
t	Thickness of the gripper	50
d_1	Shorter part of left lever arm	600
d_2	Longer part of left lever arm	1670
m_1	Shorter part of right lever arm	350
m_2	Longer part of right lever arm	1590
p	Gap between comb drive teeth	5
q	Distance of teeth overleap	30
S_1	Distance 1 of capacitive sensor	5
S_2	Distance 2 of capacitive sensor	20
N	Number of electrostatic teeth pairs	702

8.3.1 Statics Analysis

First, in order to evaluate the stiffness of the actuation part, a driving force of 150 μN is applied on the actuating movable plate in *y*-axis direction. The simulation result is shown in the Fig. 8.4. A total displacement of 23.189 μm is obtained at the left tip. Hence, the stiffness of the left arm can be computed. In addition, by generating the input displacement for the actuation plate, the amplification ratio can be calculated from the deformation result of the simulation.

Similarly, the stiffness values of K_r and K_c are calculated by applying the gripping force F_d and interaction force F_c on the right tip in *x*- and *y*-axes, and analyzing the caused displacements, respectively.

For comparison, the analytical model and FEA simulation results are summarized in Table 8.2. By taking the FEA result as the benchmark, we can observe that the errors of analytical model are less than 20%. The discrepancy is caused by the assumption made in the analytical model, in which only the bending deformations of leaf flexures are considered. The FEA result is more reliable because all kinds of deformations are taken into account in the simulation study. On the contrary, more accurate analytical models can be derived by considering all kinds of flexure deformations of the microgripper mechanism.

Additionally, the stiffness K_r of the right sensing arm is higher than the stiffness K_a of the left actuating arm as expected. This indicates that the right arm experiences a smaller deformation during the gripping operation. In addition, the interaction

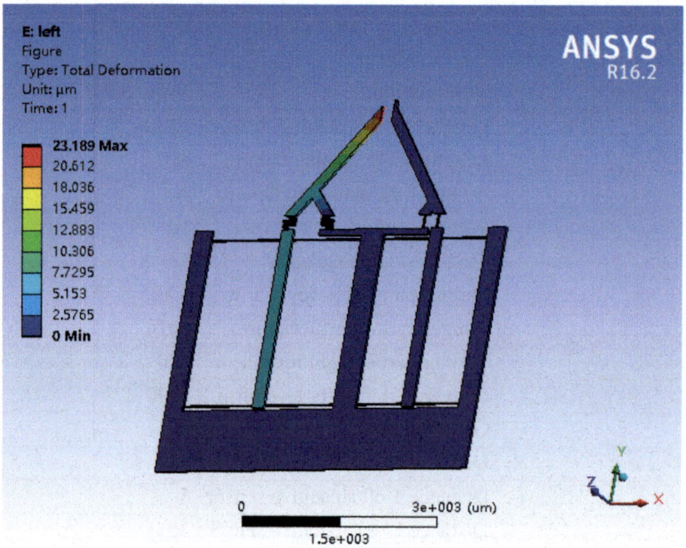

Fig. 8.4 Simulation result of deformation for the microgripper with a force of 150 μN applied at the actuating end of the movable plate

Table 8.2 Performances obtained by analytical model and FEA simulation

Performance	Analytical model result	FEA simulation result	Model error (%)
K_a (μN/μm)	2.39	2	19.5
K_r (μN/μm)	3.08	3.32	7.2
K_c (μN/μm)	62.5	61.72	1.2
R_a	2.78	3	7.3
R_s	4.54	4.31	5.3

stiffness K_c is much larger than the gripping stiffness K_r for the right arm. It means that the interaction force sensing is less sensitive than the gripping force sensing in the presented case design.

8.3.2 Cross-Axis Sensitivity Analysis

In practice, both gripping (F_d) and interaction (F_c) forces can be exerted at the microgripper. The cross-axis sensitivity analysis of the designed microgripper is conducted by performing FEA simulation study.

Unlike existing types of sensors which produce output signals in parasitic axes other than the desired working axis [17], the designed sensor creates the mixed output signal in a single axis when the forces (F_d and F_c) in two directions (x and y-axes) are applied. When a force is applied on the gripper, it causes an output displacement of the force sensor, which will be detected by the capacitive sensor and then converted into the force value. Thus, to investigate the cross-axis sensitivity of the force sensor for the microgripper, it is necessary to derive the relationship between the output signal (i.e., displacement of force sensor) and the applied gripping (F_d) and interaction (F_c) forces.

In particular, to analyze the influence of the mixed force (F_d and F_c) on the sensing of gripping force (F_d), a static structural analysis is conducted in two steps with FEA simulation. First, by applying a gripping force $F_d = 1$ μN alone, the output displacement y_d is obtained and the relationship $y_d = C_d F_d$ is derived. Hence, the parameter C_d is calculated. Second, by applying both a gripping force $F_d = 1$ μN and an interaction force $F_c = 1$ μN, the resulted output displacement y_{all} is obtained. In consideration of the relation $y_{all} = C_d F_d + C_c F_c$ and the value of C_d obtained in former step, parameter C_c is computed.

Afterward, the relationship between the induced displacement of the force sensor and the exerted mixed force can be derived as follows.

$$y_{all} = 0.3012\, F_d + 0.0162\, F_c \tag{8.22}$$

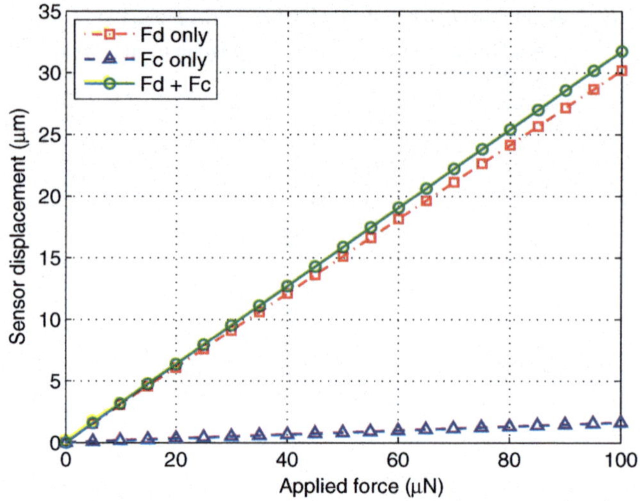

Fig. 8.5 Simulation result of cross-axis sensitivity analysis of the force sensor for gripping (F_d) and interaction (F_c) force sensing

where the displacement (y_{all}) and force (F_d and F_c) have the units of μm and μN, respectively.

Similarly, the influence of the mixed force on the sensing of interaction force (F_c) is investigated by applying F_c first and the mixed force (F_c and F_d) later. It follows that the identical expression (8.22) is obtained.

For illustrating the influence of each force component on the sensor output displacement, three curves are depicted in Fig. 8.5. It is seen that when the mixed force is applied at the gripper tip, the gripping force sensor output is increased by 5.4%, while the interaction force sensor output is increased by 18.6 times. Therefore, the mixed force causes much more influence on the interaction force sensing than that on gripping force sensing. The reason lies in that the sensor compliance in the gripping direction along x-axis is much larger than that in the interaction direction along y-axis. In the future, it is necessary to design the force sensor with similar sensitivities in the two sensing directions.

8.3.3 Dynamics Analysis

To evaluate the dynamic performance of the microgripper, modal analysis is carried out with FEA simulation study. Specifically, in order to obtain a better result, the removed masses of the actuator and capacitive sensor are added to the actuating and sensing movable plates, respectively.

The simulation results of modal analysis are shown in Fig. 8.6a and b. We can observe that the first two natural frequencies are attributed to the left actuating arm and right sensing arm at 1460.6 and 1501.6 Hz, respectively. This indicates that the first two shapes of resonant modes for the microgripper appear in the working directions of the two arms as expected.

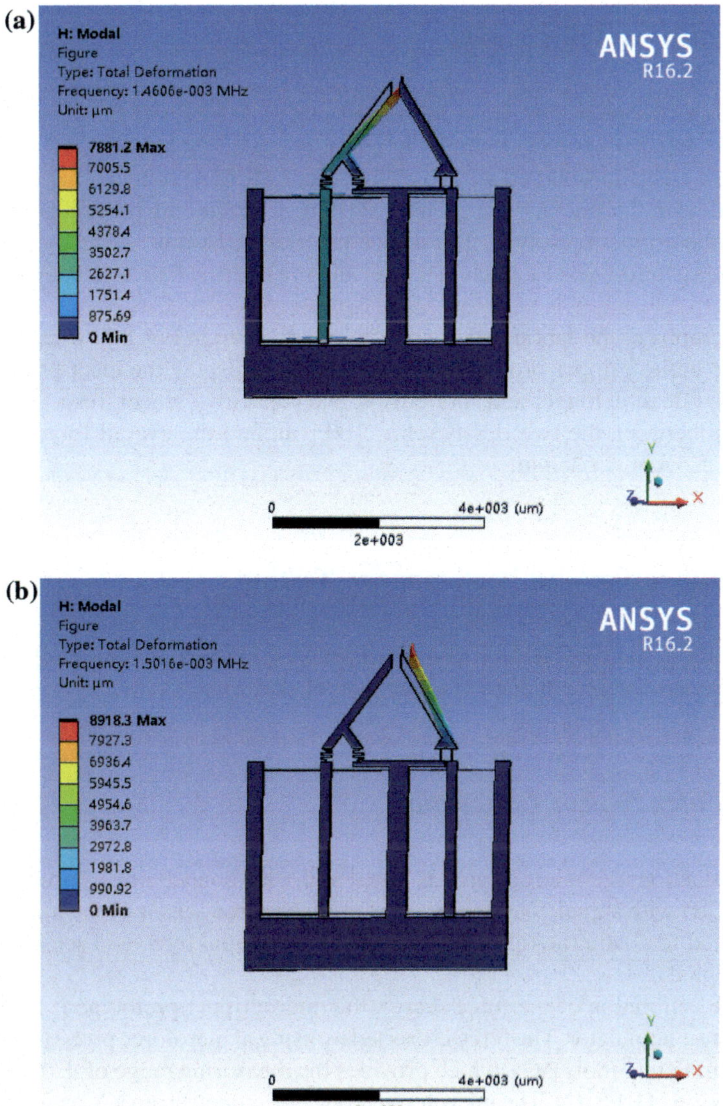

Fig. 8.6 FEA simulation results of modal analysis. **a** First resonant mode; **b** second resonant mode

Additionally, the two frequencies are relatively high, which reveals that the microgripper has a good mechanism stability and is easy to control for the future experimental studies. Besides, the resonant frequency of the right arm is larger than that of the left arm, which indicates that the right arm is stiffer than the left arm, and the right arm will experience smaller deformation during the object grasp manipulation.

8.4 Prototype Fabrication

In this work, the microgripper is fabricated by the MUMPs (Multi-Users MEMS Processes). MUMPs is a commercial program provided by MEMSCAP Inc. It provides a low cost and standard way to manufacture MEMS products.

In particular, the silicon-on-insulator (SOI) micromachining process is adopted to fabricate the gripper prototype. The device layer of SOI wafer is 25 μm, the handle layer is 400 μm for substrate manufacture, and the intermediate oxide layer is set as 2 μm.

Photographs of the fabricated microgripper are given in Fig. 8.7, where Fig. 8.7a shows the entire gripper prototype, and Fig. 8.7b, c display the electrostatic comb with folded flexure, longer and shorter tips, and capacitive sensor, respectively. The initial gap between the two tips is set as 100 μm, and the overall footprint of the gripper is 5.5 mm × 6.0 mm.

8.5 Calibration and Performance Testing

In this section, the force sensor of the gripper is calibrated and the gripper performance is examined by conducting experimental studies.

8.5.1 Force Sensor Calibration

After fabrication, the microgripper is glued and wire-bonded onto a printed circuit board (PCB) with signal conditioning circuits. The force sensor of the microgripper is calibrated to obtain the relationship between the actual force and sensor's output voltage signal.

For the calibration of the force sensor, the microgripper prototype is fixed on an XYZ micromanipulator. The force is exerted by using a microforce probe (model: FT-S100, from FemtoTools AG), which provides the measuring range of ±100 μN with the resolution of 0.05 μN. The microforce probe is mounted on an XY nanopositioner (model: P-734.2CL, from Physik Instrumente Co., Ltd.), which offers the motion range of 100 μm × 100 μm with a resolution of 0.3 nm.

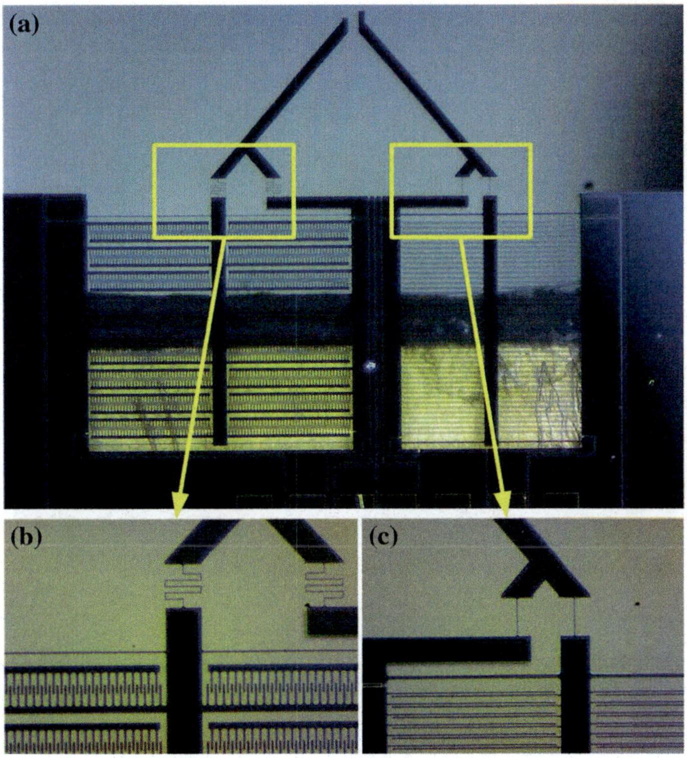

Fig. 8.7 Microscope images of the fabricated gripper prototype **a** with zoom-in view of **b** electrostatic actuator connected with folded leaf flexures and **c** capacitive sensor with straight leaf flexures

8.5.1.1 Calibration of Gripping Force Sensor

First, the calibration of the gripping force sensing is conducted. Figure 8.8 shows the experimental setup. The microgripper is fixed on an XYZ micromanipulator. It is translated toward the microforce probe which is fixed on an XY nanopositioner. This process is monitored by using an optical microscope. When the microgripper tip and microforce probe are very close to each other, the XY nanopositioner is driven by an embedded closed-loop controller to translate the sensing probe forward with a displacement of 2.5 μm at the step size of 20 nm to make a contact with the sensing tip of the gripper. Afterward, the microforce probe is translated backward in inverse direction to home position. The result is depicted in Fig. 8.9. We can observe that the output force of the microforce probe and the output voltage from the microgripper exhibit a linear relationship. The linear fit function is obtained as $y = 0.0324x - 0.0064$. The sensitivity of the force sensor is derived as the slope of the linear function, i.e., 0.0324 V/μN.

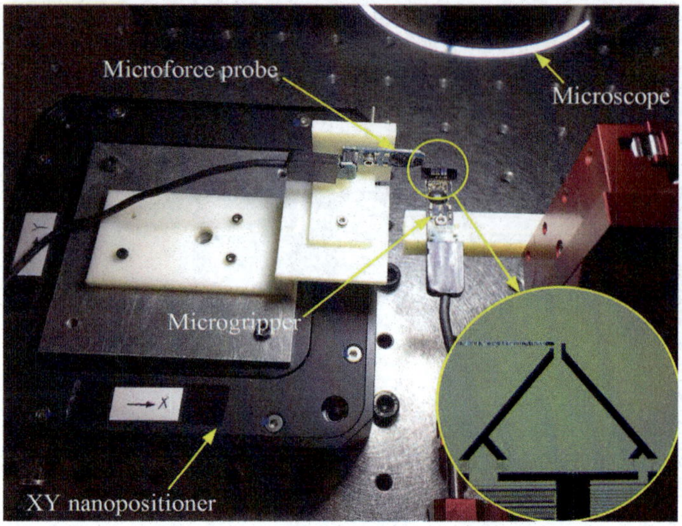

Fig. 8.8 Experimental setup for the calibration of gripping force sensing

Fig. 8.9 Calibration result of gripping force sensing for the microgripper

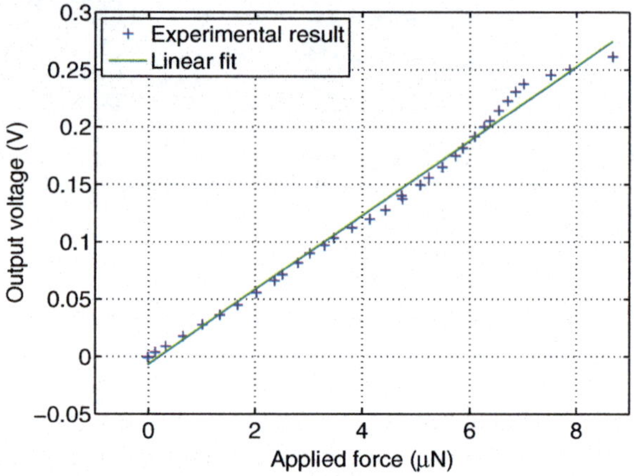

In addition, the noise signal is recorded to determine the resolution of the force sensor. Specifically, the resolution of the sensor is defined as two times standard deviation in this work, i.e., 0.58 μN. With the applied force and driven displacement, the stiffness of the right sensing arm is computed as 3.8 μN/μm. In comparison with the FEA simulation result, the experimental result is 14.5% higher. This discrepancy is mainly caused by the fabrication error.

Besides, the lever amplifier provides an amplification ratio of $R_b = 4.31$. The sensing tip delivers the maximum stroke of $\pm 6\,\mu$m. Hence, the sensing range is derived as $\pm 98.27\,\mu$N.

8.5.1.2 Calibration of Interaction Force Sensor

Second, for the calibration of the interaction force sensing for the microgripper, the experimental setup is developed as shown in Fig. 8.10. Similarly, the gripper is fixed on the XYZ micromanipulator and the microforce probe is fixed on the XY nanopositioner. Initially, the position of the XYZ micromanipulator is adjusted to translate the microgripper adjacent to the microforce probe. Then, the probe is translated by the XY nanopositioner to make a contact with the gripper tip step by step. Each step has a distance size of 10 nm. In this way, the interaction force sensing can be calibrated by recording the output voltage signal of the force sensor and output force signal of the microforce probe. The relationship between the sensing force of the probe and voltage output of the microgripper is shown in Fig. 8.11. The experimental result is fitted by a linear function $y = 0.0052x - 0.001$. The slope of the linear function reveals that the sensitivity of the interaction force sensing is 0.0052 V/μN, which is about six times lower than that of the gripping force sensing. That is, for the interaction force sensing, the sensitivity of the gripping force sensor is six times worse.

As mentioned earlier, the force sensor is designed to detect the interaction force and gripping force sequentially. Hence, the aforementioned calibrations are carried out separately.

8.5.2 Gripping Range Testing

In order to examine the gripping range, the microgripper is put under an optical microscope. When a voltage is applied to drive the left arm, the left tip of the gripper experiences a displacement. Driving by the voltage ranging from 0 to 80 V with a

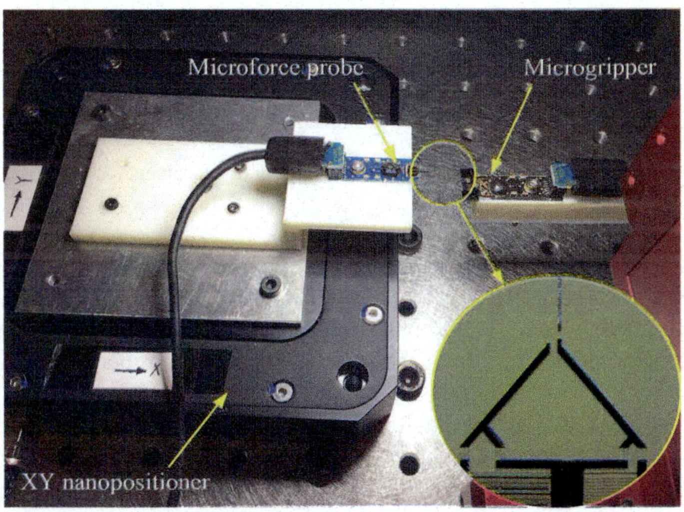

Fig. 8.10 Experimental setup for the calibration of interaction force sensing

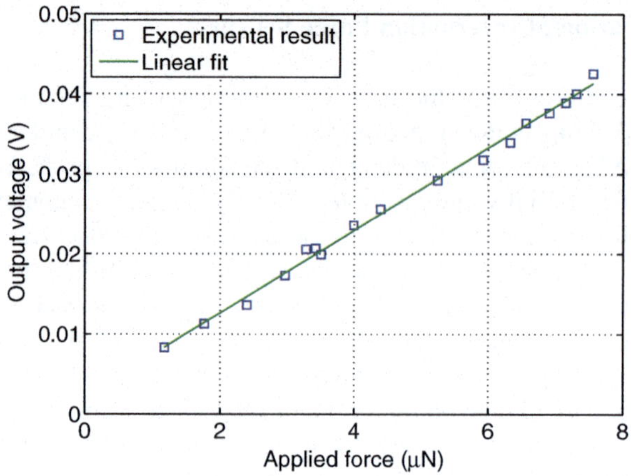

Fig. 8.11 Calibration result of interaction force sensing for the microgripper

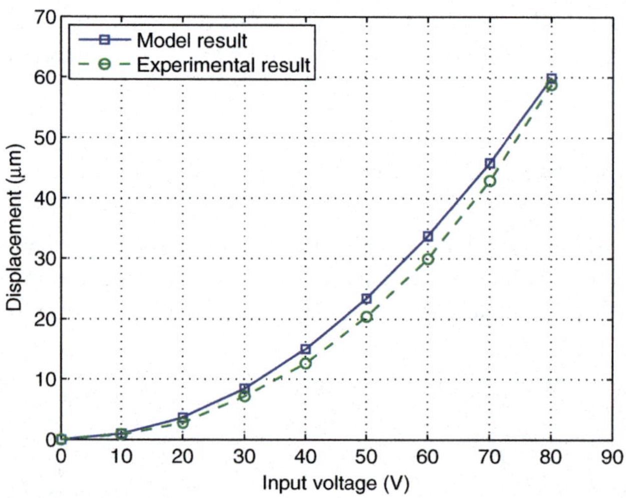

Fig. 8.12 Relationship between input voltage and gripping displacement

step size of 10 V, a series of images of the microgripper tips are acquired by using a CCD camera. The gripping range is then determined by measuring the displacement change of the left tip through image analysis technique. For comparison, the analytical model and experimental results of the relationship between the voltage and travel displacement are shown in Fig. 8.12.

We can see that the experimental result is slightly lower than the analytical model result. The error is mainly caused by the difference between the designed and fabricated parameters of the microgripper. Furthermore, to examine the side instability effect, the microgripper is driven by continuously increasing the driving voltage from 0 to 100 V. No phenomenon of side instability is observed from the experimental results. Hence, the fabricated microgripper can operate properly below the driving voltage of 100 V.

8.5.3 Bio-Cellulose Grasp Operation

To demonstrate the application of the developed microgripper, the microgripper is adopted to grasp a smooth and soft strip-shaped bio-celluose as a case study. In the experiment, the microgripper is fixed onto the XY nanopositioner, and the bio-cellulose with the diameter about 96 μm is horizontally stuck onto the XYZ micro-manipulator. Initially, the XYZ micromanipulator is adjusted manually to position the bio-sample to the middle of two tips of the microgipper. Then, the microgripper is driven by the input voltage to grasp the object and then release it.

The experimental results of the grasping and releasing procedures are illustrated in Fig. 8.13, which reveals the relationship between the gripping force and input voltage. When the driving voltage is increased to about 10 V, the sample is firmly grasped. As the driving voltage continuously increased, the soft material is squeezed to experience a plastic deformation. Hence, the sensing force exhibits a slightly nonlinear effect during the grasp-and-release operation. When the driving voltage is increased up to 75 V, the object is held for about 10 s. And then, the voltage is gradually reduced to zero. During the grasp manipulation, no side instability effect

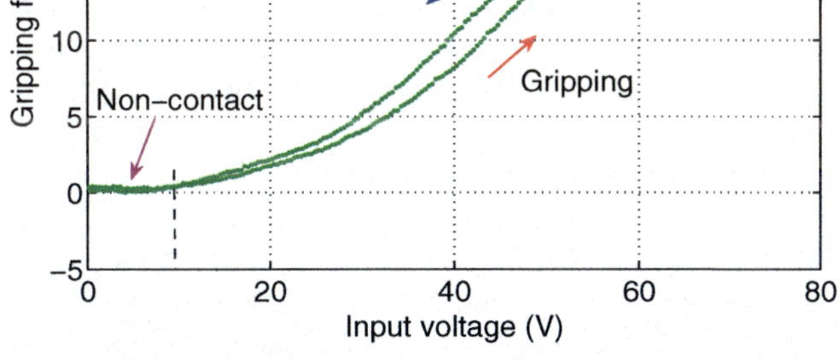

Fig. 8.13 Experimental result of the relationship between gripping force and driving voltage for grasping a bio-cellulose material

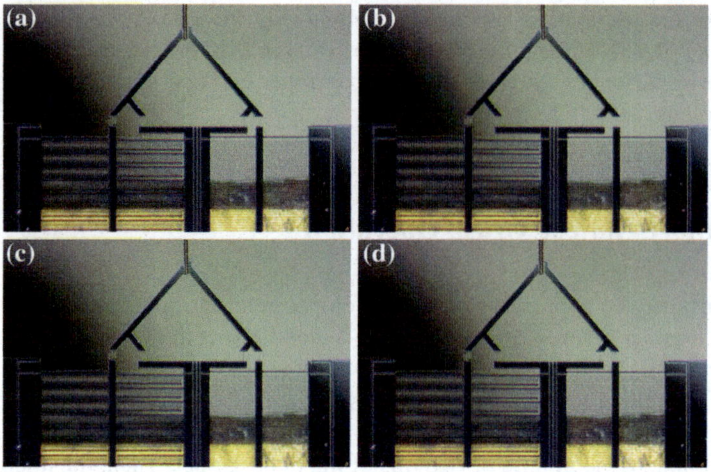

Fig. 8.14 Snapshots of grasping a bio-cellulose material of 96 μm diameter. **a** Initial state; **b** grasping; **c** grasped; **d** released

appears for the comb drive. For illustration, the snapshots of the grasp-and-release manipulation are shown in Fig. 8.14.

Furthermore, by mapping the driving voltage to displacement, the relationship between the gripping force and displacement can be generated. Then, the stiffness of the grasped sample can be calculated for material characterization for pertinent applications.

8.5.4 Further Discussion

The performance of the developed gripper is validated by the aforementioned experiments. One uniqueness of the developed microgipper lies in its force-sensing capability. It measures two directional forces, i.e., gripping and interaction force sensing, by using a single force sensor. As compared with traditional two-sensor design [12, 16], it significantly simplifies the design process, owns a more compact footprint size, and reduces the cost.

Even though the sensitivity of the interaction force sensor is six times worse, it can still work normally and meets the design goal. In the future, an optimal design of the microgripper structure will be conducted to improve the sensor sensitivity. Additionally, as compared with the previous dual-axis force-sensing gripper [21], the reported gripper exhibits a simpler mechanical structure and is much easier to manufacture. In the meantime, no side instability phenomenon is observed in the experimental studies. Besides, the gap between gripper tips can be redesigned to satisfy the gripping requirement of other sized objects.

Concerning the experimental results, a discrepancy about 10% between the experiment and analytical model is found, which is dominantly induced by the fabrication error of the etching process. A more detailed study of the sources of differences between experiments and analyses will be carried out in the future work.

8.6 Conclusions

A MEMS microgripper with integrated actuating and dual-force-sensing capabilities is designed, fabricated, and tested in this chapter. The analytical model, finite element analysis, and the experimental result exhibit a good consistency. With the voltage of 80 V, the gripper generates a gripping displacement of 59 μm. There is no side instability effect up to the driving voltage of 100 V, which exhibits a stable actuation of the microgripper. The microgripper can detect two perpendicular forces in sequence by a single sensor. The gripping force sensor owns a resolution of 0.58 μN within the force range of ±98.27 μN. The sensitivity of the interaction force sensing is six times lower than that of gripping force sensing. Moreover, the bio-sample grasp testing demonstrates the fine performance during the gripping-releasing process.

References

1. AbuZaiter, A., Nafea, M., Ali, M.S.M.: Development of a shape-memory-alloy micromanipulator based on integrated bimorph microactuators. Mechatronics **38**, 16–28 (2016)
2. Ansel, Y., Schmitz, F., Kunz, S., Gruber, H.P., Popovic, G.: Development of tools for handling and assembling microcomponents. J. Micromech. Microeng. **12**(4), 430–437 (2002)
3. Beyeler, F., Neild, A., Oberti, S., Bell, D.J., Sun, Y., Dual, J., Nelson, B.J.: Monolithically fabricated microgripper with integrated force sensor for manipulating microobjects and biological cells aligned in an ultrasonic field. J. Microelectromech. Syst. **16**(1), 7–15 (2007)
4. Chowdhury, S., Thakur, A., Wang, C., Svec, P., Losert, W., Gupta, S.K.: Automated manipulation of biological cells using gripper formations controlled by optical tweezers. IEEE Trans. Autom. Sci. Eng. **11**(2), 338–347 (2014)
5. Chronis, N., Lee, L.P.: Electrothermally activated SU-8 microgripper for single cell manipulation in solution. J. Microelectromech. Syst. **14**(4), 857–863 (2005)
6. Dechev, N., Cleghorn, W.L., Mills, J.K.: Microassembly of 3-D microstructures using a compliant, passive microgripper. J. Microelectromech. Syst. **13**(2), 176–189 (2004)
7. Estevez, P., Bank, J.M., Porta, M., Wei, J., Sarro, P.M., Tichem, M., Staufer, U.: 6 DOF force and torque sensor for micro-manipulation applications. Sens. Actuators A, Phys. **186**, 86–93 (2012)
8. Grade, J.D., Jerman, H., Kenny, T.W.: Design of large deflection electrostatic actuators. J. Microelectromech. Syst. **12**(3), 335–343 (2003)
9. Howell, L.L.: Compliant Mechanisms. John Wiley & Sons (2001)
10. Khan, F., Bazaz, S.A., Sohail, M.: Design, implementation and testing of electrostatic SOI-MUMPs based microgripper. Microsyst. Technol. **16**(11), 1965–1967 (2010)
11. Kim, D.H., Lee, M.G., Kim, B., Sun, Y.: A superelastic alloy microgripper with embedded electromagnetic actuators and piezoelectric force sensors: a numerical and experimental study. Smart Mater. Struct. **14**(6), 1265–1272 (2005)

12. Kim, K., Liu, X., Zhang, Y., Sun, Y.: Nanonewton force-controlled manipulation of biological cells using a monolithic MEMS microgripper with two-axis force feedback. J. Micromech. Microeng. **18**(5), 055,013 (2008)
13. Krijnen, B., Brouwer, D.M.: Flexures for large stroke electrostatic actuation in MEMS. J. Micromech. Microeng. **24**(1), 015,006 (2014)
14. Molhave, K., Hansen, O.: Electro-thermally actuated microgrippers with integrated force-feedback. J. Micromech. Microeng. **15**(6), 1265–1270 (2005)
15. Piriyanont, B., Moheimani, S.O.R.: MEMS rotary microgripper with integrated electrothermal force sensor. J. Microelectromech. Syst. **23**(6), 1249–1251 (2014)
16. Qu, J., Zhang, W., Jung, A., Sliva, S., Liu, X.: A MEMS microgripper with two-axis actuators and force sensors for microscale mechanical characterization of soft materials. In: Proc. IEEE Int. Conf. on Automation Science and Engineering (CASE), pp. 1620–1625 (2015)
17. Sankar, A.R., Das, S., Lahiri, S.K.: Cross-axis sensitivity reduction of a silicon MEMS piezoresistive accelerometer. Microsyst. Technol. **15**(4), 511–518 (2009)
18. Solano, B., Wood, D.: Design and testing of a polymeric microgripper for cell manipulation. Microelectro. Eng. **84**(5), 1219–1222 (2007)
19. Wei, J., Porta, M., Tichem, M., Sarro, P.M.: A contact position detection and interaction force monitoring sensor for micro-assembly applications. In: Proc. 15th Int. Conf. Solid-State Sens., Actuators Microsyst. (TRANSDUCERS), pp. 2385–2388 (2009)
20. Xu, Q.: Adaptive discrete-time sliding mode impedance control of a piezoelectric microgripper. IEEE Trans. Robot. **29**(3), 663–673 (2013)
21. Xu, Q.: Design, fabrication, and testing of an MEMS microgripper with dual-axis force sensor. IEEE Sensors J. **15**(10), 6017–6026 (2015)
22. Yang, S., Xu, Q.: Design of a microelectromechanical systems microgripper with integrated electrothermal actuator and force sensor. Int. J. Adv. Robot. Sys. **13**(5), 1729881416663,375 (2016)
23. Yang, S., Xu, Q., Nan, Z.: Design and development of a dual-axis force sensing MEMS microgripper. Journal of Mechanisms and Robotics **9**(6), 061,011 (2017)
24. Zhang, R., Chu, J., Wang, H., Chen, Z.: A multipurpose electrothermal microgripper for biological micro-manipulation. Microsyst. Technol. **19**(1), 89–97 (2013)

Chapter 9
Design, Analysis, and Development of a Piezoelectric Microsyringe Pump

Abstract This chapter presents the design, analysis, and testing of a microsyringe pump for delivering the liquid inside a tiny tube precisely. To facilitate precision actuation of the microsyringe, a uniaxial micromotion pump is devised with flexure-based compliant mechanisms. Based on the specifications on travel range and load capability, a compliant stage with translational motion is developed. Parameter optimization of the stage mechanism is conducted to meet the pump performance requirements. The pump performance is validated by performing simulation study with finite element analysis (FEA). In addition, a prototype pump is fabricated by using a piezoelectric actuator. A PID position feedback controller is implemented, and experimental investigations are carried out for performance test of the precision pump with the help of a microscope.

9.1 Introduction

Based on microelectromechanical systems (MEMS) technique, micropumps can be manufactured in smaller size along with higher precision and performance [2]. They have been applied extensively in a vast of domains [10]. Typical applications of the micropumps involve drug delivery, chemical analysis, DNA analysis, environmental monitoring.

Currently, there are a number of medical micropumps which are driven by different pumping actuation and mechanisms. Each type of micropump has its own advantages and disadvantages. Typically, a medical micropump device consists of the micropump, actuator, drug reservoir, microsensors, and valves, that can deliver drug from the drug reservoir to the body [8]. Yet, as the drug reservoir is specially fabricated, the role of the device is constrained to the delivery of a single type of liquid (drug). It is difficult to change different liquids via drug reservoir. In order to overcome such issue, the development of a microsyringe pump is conducted in this chapter.

For the micropump development, various drivers have been adopted to actuate the pump [7, 13]. In particular, piezoelectric actuator (PEA) works on the basis of inverse piezoelectric effect of piezoelectric ceramics [6]. As compared with other

Q. Xu, *Micromachines for Biological Micromanipulation*,
https://doi.org/10.1007/978-3-319-74621-0_9

types of micropump actuation approaches such as magnetostrictive alloy and shape memory alloy, PEA exhibits several attractive features. For instance, PEA provides fast response, large blocking force, and high energy efficiency. It plays a crucial role in the area of smart actuator systems. Its dominant applications involve precision motors, positioners, and vibration resonators.

At the same time, microsyringe pump has been broadly applied in basic scientific studies including the quantitative analysis of biology and chemistry [4, 5]. Nowadays, majority of the commercial microsyringe pumps at the market work based on mechanical transmission such as worm drive, that can obtain continuous and sufficiently stable output, whereas the backlash in mechanical components is difficult to avoid [11, 12]. To this end, the conceptual design of a microsyringe pump with flexure mechanism is reported in this chapter to overcome the current issues.

The microsyringe pump device consists of a PEA, a displacement amplifier, a compound parallelogram flexure (CPF) mechanism, some fixtures, and a syringe connected to the tiny tube. To facilitate the testing with rapid response, PEA and capacitive displacement sensor are employed as actuator and sensor, respectively. To generate the desired performance and to constrain the physical dimension of the device, finite element analysis (FEA) simulation and parametric optimization are conducted in the design procedure. For practical application of microsyringe pump, a proportional–integral–derivative (PID) control is implemented in the control process. The outlet of the syringe is connected to a tiny tube, and the liquid level is monitored with a microscope. Experimental study is carried out to verify the concept design and to characterize the performance of the fabricated microsyringe pump.

9.2 Mechanism Design

The mechanism design of the flexure-based microsyringe pump is presented in this section.

9.2.1 Design of Displacement Amplifier

In order to overcome the limited travel stroke of PEA actuator, a displacement amplifier is adopted to magnify the PEZ stroke so as to generate a large output displacement [16].

For the calculation of the amplification ratio for the amplifier (Fig. 9.1), we assume that each of the flexure hinge has a one degree-of-freedom (1-DOF) rotational compliance and other components are rigid bodies. The rotational DOF of flexure hinge arises from its rotational deformation. Figure 9.2 depicts a quarter model of the displacement amplifier. Based on geometric relation analysis, the amplification ratio of the bridge-type amplifier can be calculated as follows [15].

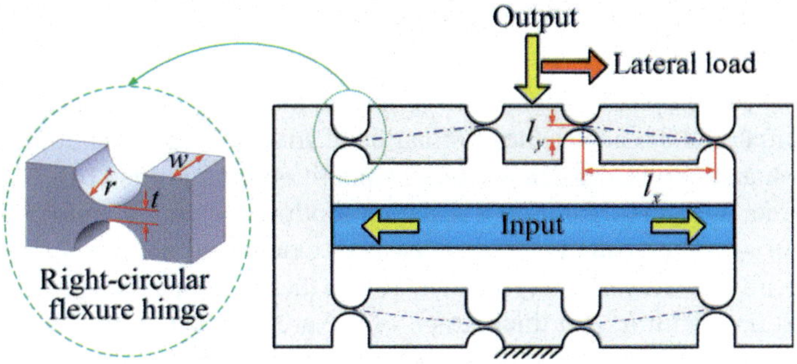

Fig. 9.1 A flexure-based bridge type of displacement amplifier with backward output direction

$$A = \frac{\Delta y}{\Delta x} = \frac{l_a \sin\alpha - \sqrt{l_a^2 \sin^2\alpha - \Delta x^2 - 2l_a \Delta x \cos\alpha}}{\Delta x} \tag{9.1}$$

where α is the initial inclined angle and l_a is the length of the beam between two flexure hinges.

According to Eq. (9.1), given the input displacement Δx, the amplification ratio is governed by l_a and α. Referring to Fig. 9.1, we observe that the length l_a and the angle α are related to the radius r of hinge, the thickness t of hinge, the block length l_x, and the block width l_y. Thus, we only need to consider four design variables in the design procedure, i.e.,

$$\text{Amplifier}(r, t, l_x, l_y). \tag{9.2}$$

Fig. 9.2 One-quarter model of the compound bridge-type amplifier with parameters

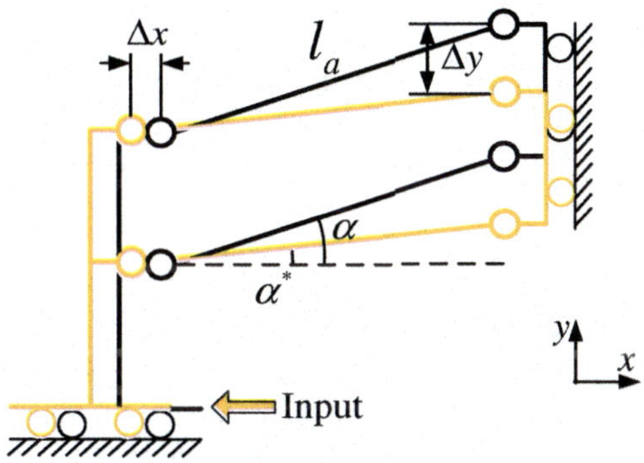

9.2.2 Design of Parallelogram Flexure

A flexure-based compliant mechanism produces motion by using elastic deformations of the material. It enables the elimination of adverse effects including clearance and backlash and allows the generation of smooth output motion [14].

To generate a large-stroke pure translation motion, the compound parallelogram flexures (CPFs) as shown in Fig. 9.3a, b have been extensively adopted in the literature [1, 9]. Evidently, the flexures play a crucial role in the entire mechanism. In particular, the length l, the width h, and thickness b of flexures are the main design variables which affect the mechanism performance.

$$\mathrm{CPF}(l, h, b) \tag{9.3}$$

Figure 9.3 depicts two different types of CPFs. Correspondingly, there are two design solutions using the two CPFs. Figure 9.4a, b show the two possible mechanical designs of the compliant mechanism.

In order to select a suitable solution from the two designs, different issues including the compactness, modifiability, and material weight of design are taken into account.

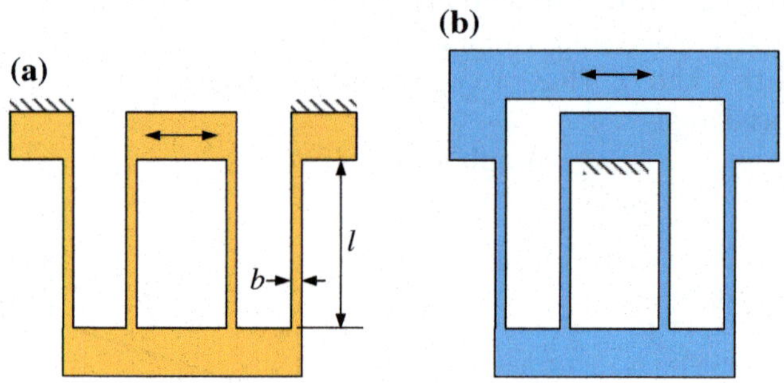

Fig. 9.3 Illustrations of two different compound parallelogram flexures (CPFs)

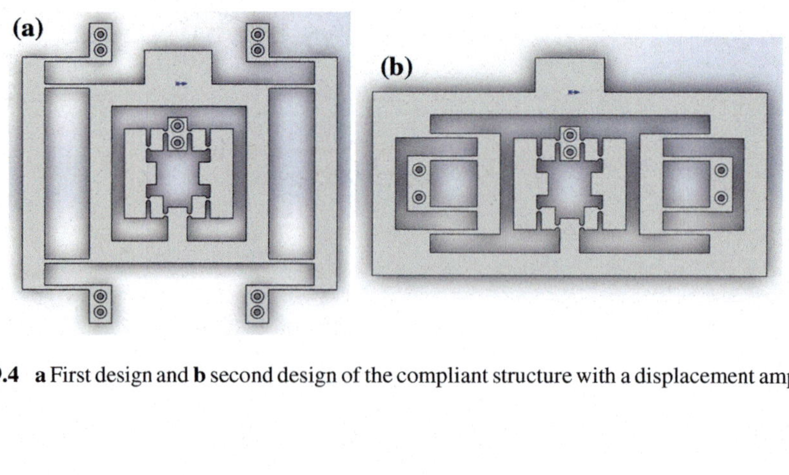

Fig. 9.4 **a** First design and **b** second design of the compliant structure with a displacement amplifier

Considering that the second design occupies a larger footprint and material weight, the first design is chosen in this work.

9.3 Optimization Design and Simulation Study

In this section, the optimal design of the mechanism parameters is conducted and simulation study is carried out to verify the mechanism performance.

9.3.1 *Optimization Setup*

Considering the requirement on dynamics characteristics for a flexure mechanism, the natural frequency in the working direction is an important indicator to reflect the performance of the mechanism. Therefore, the objective for the optimization is to generate the highest natural frequency under the constraints imposed by the input stiffness, output displacement, and the safety factor of the material for satisfying the minimum requirement.

Referring to Eqs. (9.2) and (9.3), the design variables of main structure can be expressed as follows.

$$\text{Main structure}(r, t, l_x, l_y, l, h, b) \tag{9.4}$$

To facilitate the manufacturing process, the thickness h of the whole mechanism is chosen by prior experience. In particular, two values are chosen for the thickness, i.e., 6 and 10 mm. Hence, the number of design variables is reduced to six as follows.

$$\text{Main structure}(r, t, l_x, l_y, l, b) \tag{9.5}$$

In order to achieve a compact size for the main structure, reasonable ranges are assigned for these six variables. Hence, an optimization problem with a single objective and multiple constraints is constructed.

9.3.2 *Optimization Results*

Commonly, analytical models are required to solve the optimization problem. Considering that a complete analytical modeling is time-consuming and the model error always exists to some extent, the optimization is implemented by using the built-in optimization module in ANSYS Workbench simulation software of finite element analysis. In this way, the analytical models are not required. Instead, only the objective and constraints need to be specified in the software.

Table 9.1 Constraints on the main structure performance

Parameter	Constraint
Output displacement	Greater than 7x input displacement
Safety factor	Inside the range [2, 5]
Thickness of positive-stiffness beam	4 mm
Input stiffness	Smaller than actuator stiffness
Natural frequency	Greater than 100 Hz

Table 9.2 Constraints on the main design variables

Design variable	Symbol	Range (mm)
Radius of right-circular hinge	r	[0.5, 2.5]
Minimum width of right-circular hinge	t	[0.2, 1]
Length of block between two hinges	l_x	[10, 25]
Width of block between two hinges	l_y	[3, 15]
Length of leaf flexure	l	[10, 25]
Width of leaf flexure	b	[0.2, 1]

In addition, the adaptive single-objective (ASO) algorithm is a gradient-based approach to generate a refined, global optimization result [3]. To solve the optimization problem with a single objective and multiple constraints, the ASO approach is adopted in the optimization process. The objective and constraints are shown in Table 9.1, and the parameter ranges for each design variable are depicted in Table 9.2. The optimization module automatically chooses and ranks the best results. The optimal results of the design with two different thicknesses values are given in Table 9.3.

Additionally, the material is assigned as Al-7075 alloy in the optimization process. In the case design, the stiffness of the adopted PEA is 2.3×10^8 N/m, that is one order of magnitude higher than the input stiffness of each mechanism design. Thus, the actuator's stiffness is neglected in the design procedure.

By comparing the objective function values of the designs, it is found that Design 1 with 10-mm thickness provides the highest natural frequency. Therefore, the design variable values of Design #1 with $h = 10$ mm is chosen. To facilitate the fabrication process, the design variable values are truncated to take two decimal numbers.

9.3.3 Simulation Results

As the parameter values take two decimal values for simplicity, the simulation study with finite element analysis (FEA) is conducted again to verify the performance of the flexure mechanism before the prototype fabrication.

Table 9.3 Optimization results of structure design with two different thicknesses values

Design variables	Thickness h = 10 mm		Thickness h = 6 mm	
	Design #1	Design #2	Design #1	Design #2
l (mm)	17.89	18.797	15.579	14.995
b (mm)	0.65854	0.68016	0.6513	0.64723
t (mm)	0.52904	0.5558	0.452	0.43517
r (mm)	1.0847	1.11395	0.92815	0.9789
l_x (mm)	11.826	12.324	15.472	15.73
l_y (mm)	3.7624	3.8709	3.3698	3.481
Output displacement (μm)	101.98	102.37	103.45	102.33
Safety factor	4.4558	4.5345	4.8615	4.9697
Input stiffness (N/m)	32258064.52	32739654.27	20218358.27	20403166.57
Natural frequency (Hz)	242.96	240.24	221.56	220.59

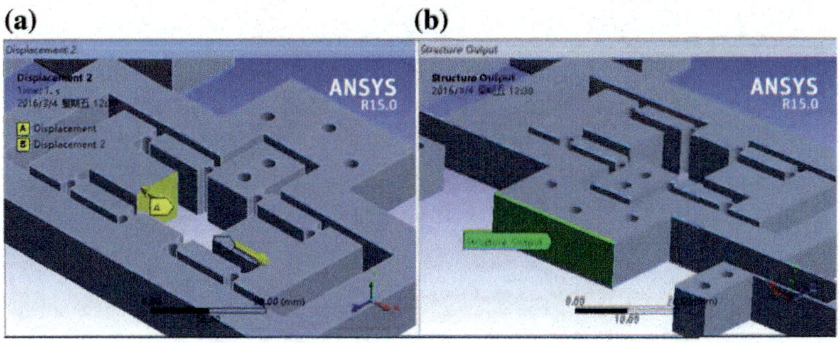

Fig. 9.5 Input and output of the simulation study. **a** The two displacement inputs; **b** the target output displacement

Using the static structural analysis module in ANSYS Workbench, the simulation is performed to generate the maximum displacement. In this work, the adopted PEA provides the maximum output stroke of 14.5 μm. As depicted in Fig. 9.5a, an input displacement of 7.25 μm is applied at each input side of the displacement amplifier. The output displacement of the whole mechanism is measured as the target.

The simulation result for the output displacement is 102.18 μm, which is more than seven times larger than in the input displacement. It fulfills the output motion requirement. With this input, a safety factor of 4.5017 is achieved, that indicates that the material works in the safe range, as shown in Fig. 9.6.

Furthermore, a modal analysis is conducted to evaluate the dynamics performance. The simulation results of the first six resonant modes are shown in Fig. 9.7. It is observed that the first natural frequency occurs at 240.57 Hz and the corresponding resonant mode is translation along the working direction.

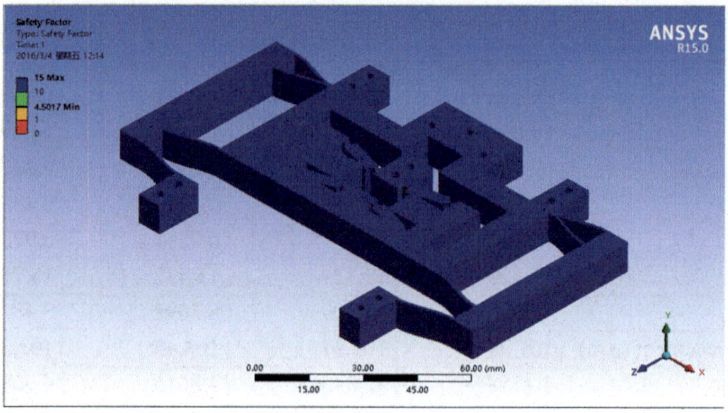

Fig. 9.6 Simulation result of the safety factor

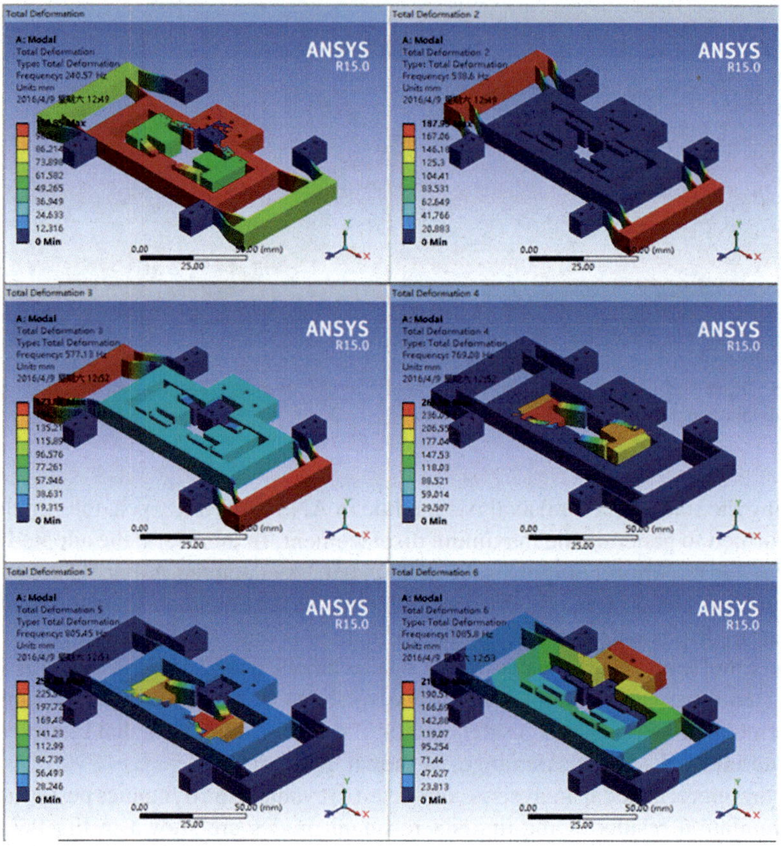

Fig. 9.7 Simulation results of the first six resonant modes of the structure

9.4 Prototype Development and Experimental Results

In this section, a prototype of the designed microsyringe pump is fabricated and experimental study is conducted to test its performance.

9.4.1 Prototype Fabrication and Assembly

Figurc 9.8 shows a CAD model of the microsyringe pump which is developed with optimal parameters. As depicted in Fig. 9.9, the prototype is fabricated with a plate of Al-7075 material by wire EDM process with the tolerance of 0.01 mm. In order to decrease the fabrication cost and the entire weight of the mechanism, the fixtures for the sensor and syringe are fabricated with polymethyl methacrylate (PMMA) material with the tolerance of 0.1 mm. This does not influence the output motion of the syringe pump while guaranteeing the compactness of the design.

The entire experimental setup is shown in Fig. 9.10. The microsyringe is actuated by a PEA (model: TS18-H5-202, from Piezo Systems, Inc.). The PEA is driven through a high-voltage amplifier (model: EPA-104, from Piezo System, Inc.). The

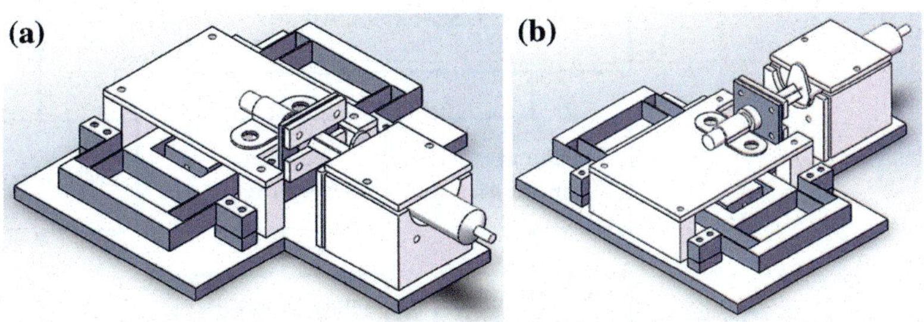

Fig. 9.8 CAD model of the microsyringe system. **a** and **b** are different views

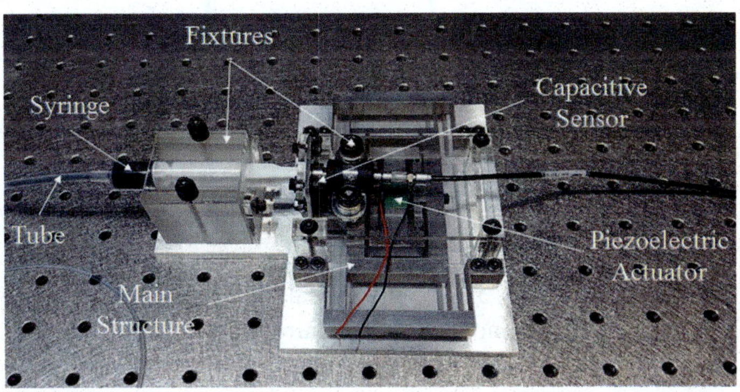

Fig. 9.9 The fabricated prototype of the microsyringe pump

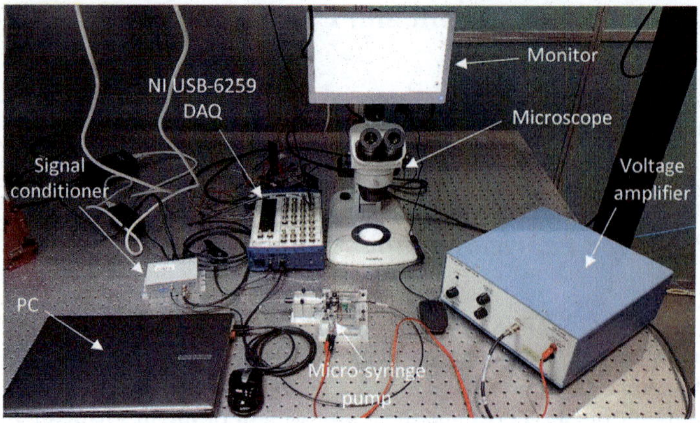

Fig. 9.10 Experimental setup of the microsyringe pump system

Fig. 9.11 Illustration of hardware connection scheme

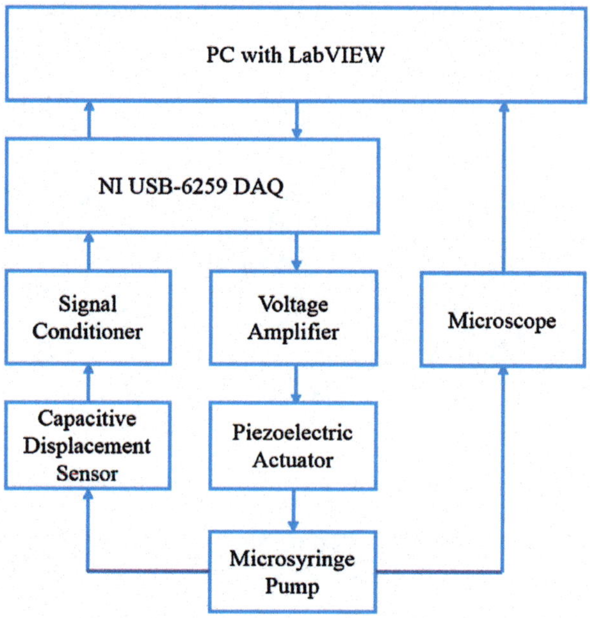

output displacement is measured by a capacitive displacement sensor (model: D-510.050, from Physik Instrumente Co., Ltd.). The control algorithm is realized by a personal computer with a data acquisition board of NI USB-6259 (from National Instruments Corp.). In addition, an upright optical microscope (model: SZX16, from Olympus Corp.) is adopted to monitor the operation of the microsyringe pump. The hardware connection is illustrated in Fig. 9.11.

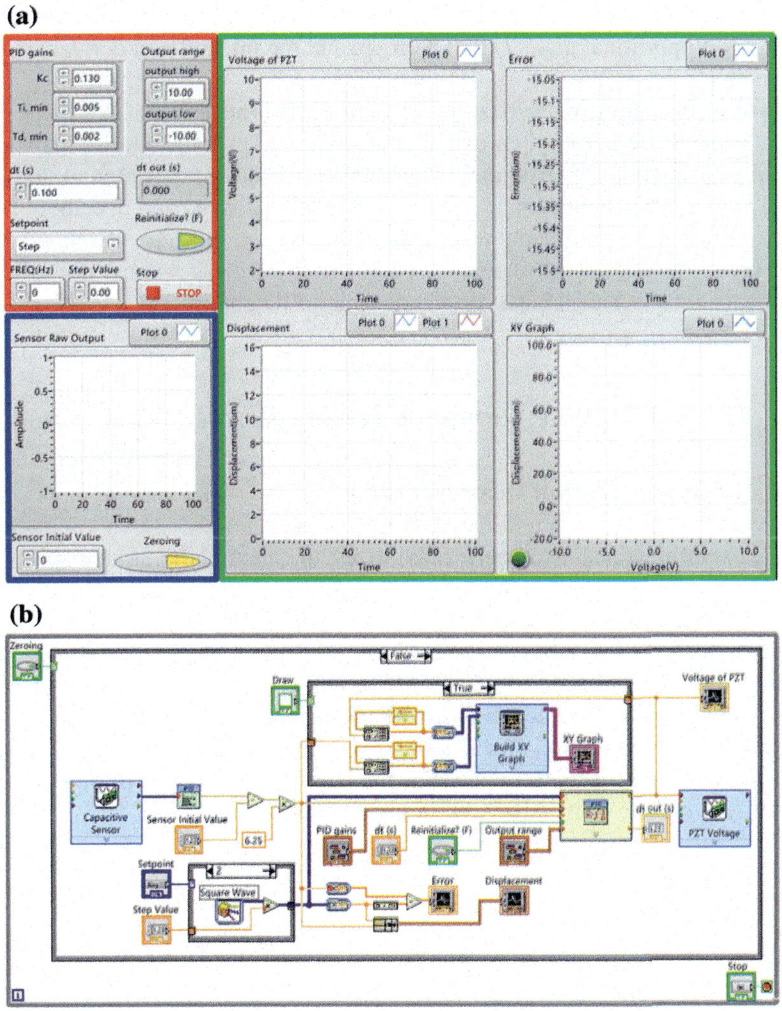

Fig. 9.12 **a** GUI and **b** control block diagrams of the control scheme developed using LabVIEW software

9.4.2 Controller Setup

The graphical user interface (GUI) of the control algorithm is shown in Fig. 9.12. For a precise motion control of the microsyringe pump, a PID controller is implemented with LabVIEW in this work. The PID control algorithm is given below.

$$u(t) = K_c\left[e(t) + \frac{1}{T_i}\int_0^t e(t)\mathrm{d}\tau + T_d\frac{\mathrm{d}e(t)}{\mathrm{d}t}\right] \tag{9.6}$$

where t is the time variable, $u(t)$ is the control action, $e(t)$ denotes the control error, K_c is proportional gain, T_i and T_d represent the integral and derivative times, respectively.

Regarding the tuning of the PID control gains, there is no universal approach which can treat every system with their own characteristics. Thus, the control gains of the PID controller are tuned by trial-and-error approach through experimental study. The purpose is to make a compromise between the response speed and overshoot for the response. With the experimental testing and comparison, the gains are finally tuned as $K_c = 0.140$, $T_i = 0.004$ (min), and $T_d = 0.001$ (min), with $dt = 0.1$ s. The following experiments are conducted based on the designed PID controller.

9.4.3 *Microsyringe Performance Testing Results*

9.4.3.1 Maximum Output Displacement

First, the experimental study is performed to test the maximum output displacement of the microsyringe. Specifically, the PEA is driven by a 0.1 Hz sinusoidal signal with the amplitude of 5 V. The input voltage is amplified by ten times via the voltage amplifier, which is then applied to the PEA. The experimental result is depicted in Fig. 9.13.

We can observe that the actual output displacement is 81.9 μm, which is 19.8% lower than the simulation result (102.18 μm). The discrepancy between the actual and simulation results is mainly caused by the fabrication error and preloading effect of the PEA, which are not considered in the simulation study.

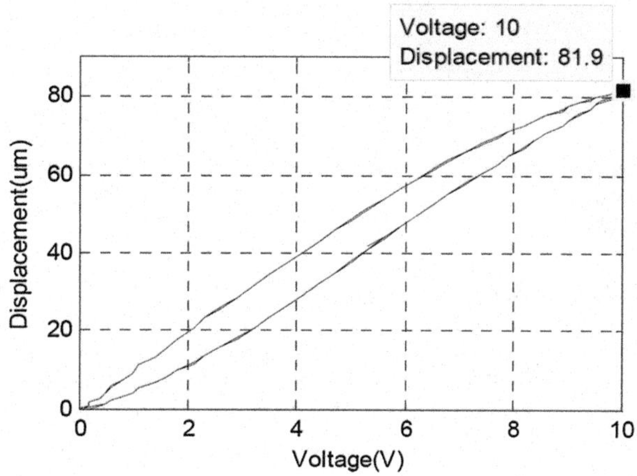

Fig. 9.13 Relationship between the output displacement and input voltage for the microsyringe

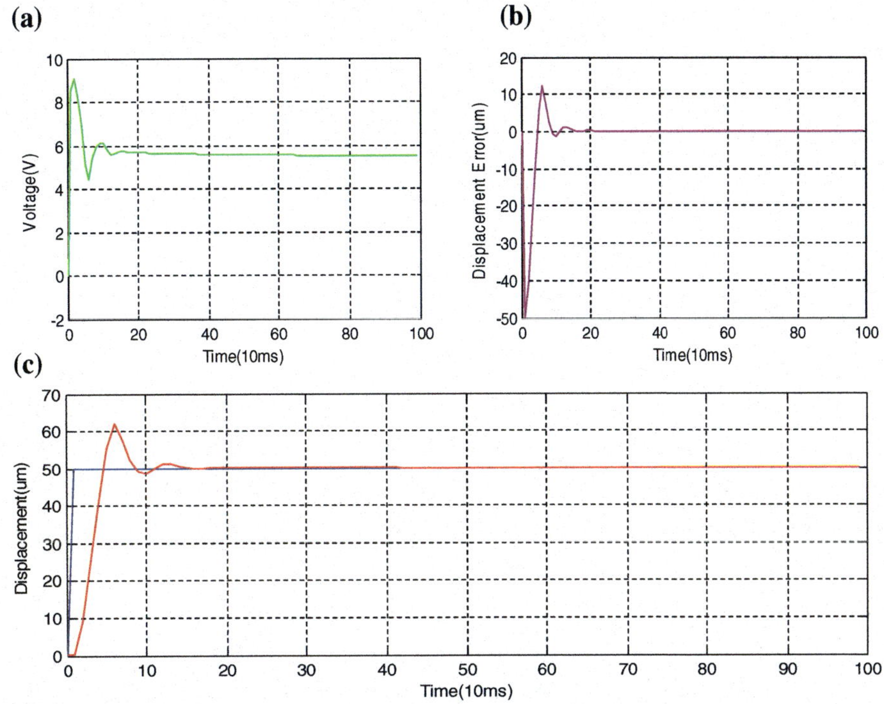

Fig. 9.14 Experimental results of step response. **a** The control action; **b** displacement error; **c** time history of the displacement

9.4.3.2 Dynamic Performance

Second, the dynamics characteristic of this system is examined by performing a step response test and two sinusoidal motion tracking tests with different frequencies.

First, in the step response experiment, a step signal varying from 0 to 50 μm is used. Figure 9.14 shows the experimental results including the driving voltage, output displacement, and displacement error.

Next, the sinusoidal response experimental is executed by commanding sinusoidal reference inputs to the controller. By adopting the same amplitude of 40 μm with an offset of 40 μm, the frequency of the sinusoidal signal is changed from 0.5 Hz to higher value. Figure 9.15 shows the tracking results of 0.5 Hz sinusoidal reference input motion.

By comparing Fig. 9.15 with the results generated in higher frequencies, we can find that the system works well in low frequency, whereas the performance at high frequency is not good enough. The time delay is more obvious at higher frequency, that causes a larger error and more evident hysteresis effect. This is the limitation of PID controller with constant control gains. Further experiments indicate that the control performance is better if the gains are readjusted for high-frequency tracking,

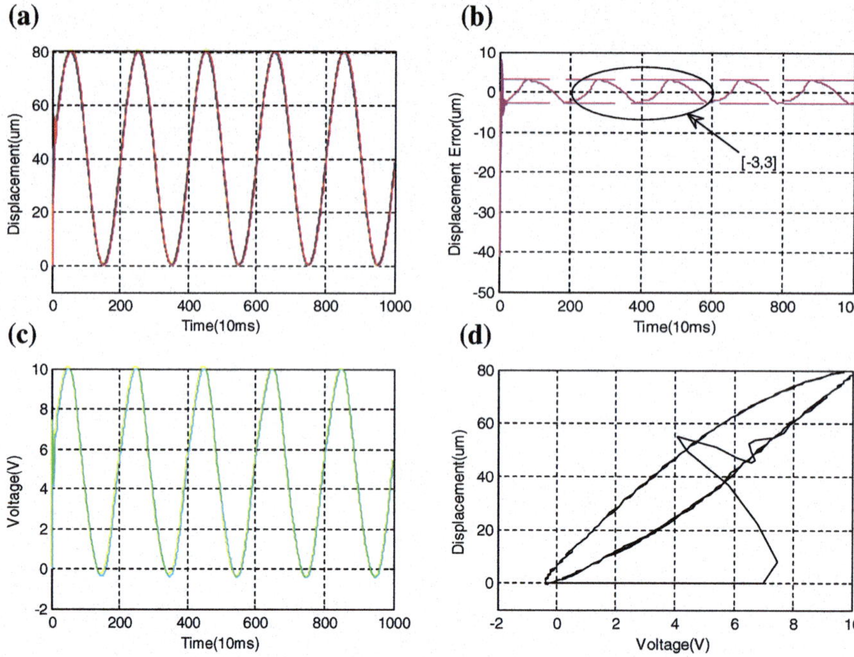

Fig. 9.15 Experimental results of 0.5 Hz sinusoidal motion tracking. **a** Time history of the displacement; **b** displacement error; **c** control action; **d** displacement-voltage hysteresis loop

whereas the performance is worse in other situations. To overcome the problem, an advanced control algorithm is required, which will be conducted in the future work.

9.4.3.3 Resolution Testing

Third, the resolution of the microsyringe pump system is tested by experiment. It is important to determine the minimum output which can be reached by the system. The experiment is performed by gradually decreasing the amplitude of a square-wave reference input, until the lowest amplitude can still be discriminated in terms of the peak and valley. The experimental results are shown in Fig. 9.16, which indicates that the system resolution is around 5 nm.

9.4.4 Microsyringe Pump Performance Testing Results

The foregoing experimental results are obtained by the capacitive displacement sensor, which only represent the performance for the main structure of the microsyringe.

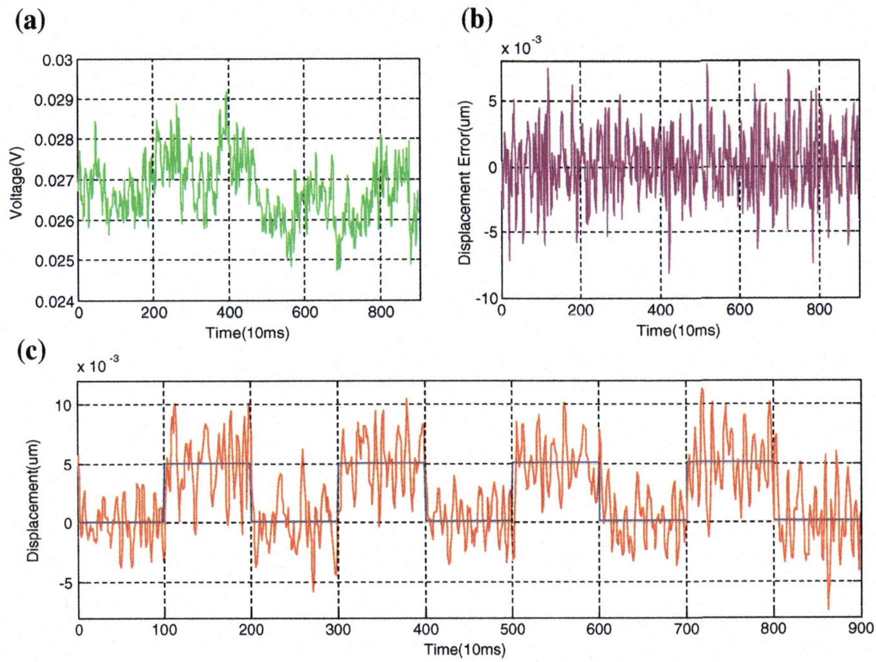

Fig. 9.16 Resolution testing results. **a** The control action; **b** displacement error; **c** time history of the displacement

Fig. 9.17 Illustration of liquid parameters inside a small tube

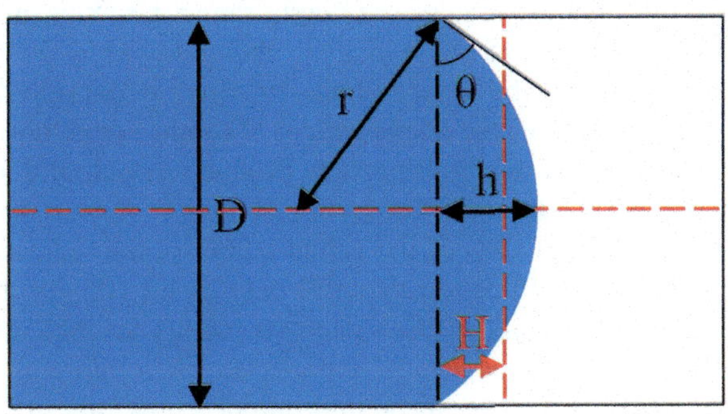

In order to confirm the actual performance of the microsyringe pump, a microscope is adopted to monitor the actual pump output inside a tiny tube.

9.4.4.1 Liquid Level Identification

Due to the capillarity effect inside a tiny tube, the actual liquid level cannot be measured directly. Thus, an analysis is required to determine the liquid level.

Suppose that the liquid surface exhibits a spherical shape. The related parameters can be directly measured from the microscope image by image processing. As shown in Fig. 9.17, the diameter D of the tube, the radius r of liquid surface, the contact angle θ, and the height h of spherical surface can be determined. Afterward, the actual height H of a convex surface is computed as follows.

$$H = \frac{4}{D^2} \int_0^h [(h - y)(D \cos \theta - h + y)]dy \tag{9.7}$$

In the case of a concave surface, Eq. (9.7) can be modified as follows.

$$H = h - \frac{4}{D^2} \int_0^h [(h - y)(D \cos \theta - h + y)]dy. \tag{9.8}$$

9.4.4.2 Syringe Output Volume Calculation

Next, the output volume of the microsyringe pump is determined by experiments conducted with the microscope. In particular, a 0.1 Hz sinusoidal signal with the amplitude of 5 V and offset of 5 V is employed to actuate the PEA via the high-voltage amplifier.

The inner radius of the tube is $D = 0.8$ mm. Other related parameters are shown in Fig. 9.18. The vertical lines in Fig. 9.18d indicate the actual liquid level of the "valley" and "peak" shown in Fig. 9.18a, b, respectively. The distance between the two lines is measured as the output range, i.e., 8.64 mm. The data are measured by analyzing Fig. 9.18a, b, which are shown in Table 9.4.

In addition, the inner radius of syringe is measured as 12.5 mm. Therefore, the syringe amplifies the displacement of liquid in the tube by about 244.14 times. This means that the microsyringe can produce an output displacement of 244.14×81.9 μm = 20.00 mm in the liquid tube.

However, in view of the actual output of the main mechanism and the tiny tube, the actual amplification factor is only about 105.49, that is much smaller than the calculated value. The discrepancy is mainly attributed to the syringe. In reality, the rubber part of the syringe imposes an extra damping to the system, which has a remarkable effect on the liquid output in the tube. Due to the deformation of the rubber part of the syringe, the output in tube is much smaller than the computational value.

Moreover, it is desirable to calculate the minimum volume of the liquid that can be delivered by the microsyringe pump. Suppose that the syringe is a linear device, the minimum volume can be expressed as follows.

$$V_{\min} = \text{Resolution} \times \text{Amplification factor} \times \text{Tube sectional area} \tag{9.9}$$

In consideration of the 5-nm resolution of the main mechanism and the amplification factor of 105.49 for the microsyringe, the minimum volume is estimated as 0.265 nL.

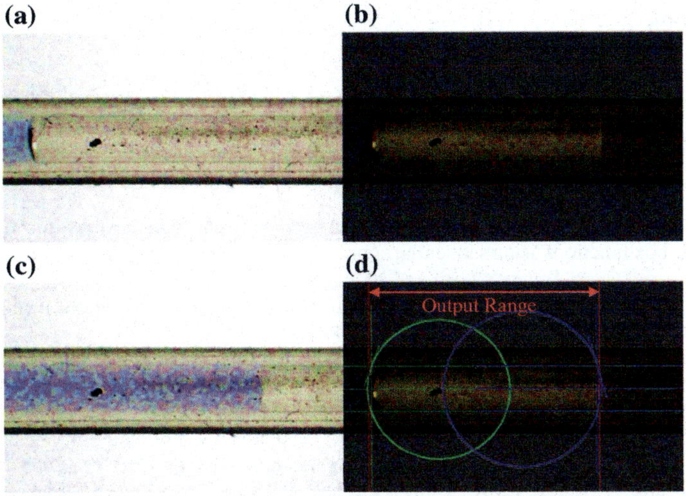

Fig. 9.18 Image processing for liquid volume calculation. **a** The "valley" level; **b** the "peak" level; **c** merged image; **d** analysis of the merged image

Table 9.4 Image analysis results of the liquid levels

Parameters	"Valley" level	"Peak" level
θ (degree)	19.5	16.7
D (mm)	0.8	0.8
h (mm)	0.074419	0.060465
H (mm)	0.0338	0.031349
Output range (mm)	8.64	

9.5 Conclusion

This chapter presents the design, optimization, and control of a flexure-based microsyringe pump. Based on a parameterized CAD model, the parameter optimization is conducted to achieve the highest natural frequency. A simulation study with FEA is performed to verify the optimal design and to examine the mechanism performance. A prototype microsyringe is developed for experimental study. To facilitate the calculation of the output displacement for the microsyringe pump, the expression for the actual liquid height is derived. Results show that the microsyringe pump is able to deliver the minimum volume of 0.265 nL. In the future, an advanced control algorithm will be realized to enhance the performance of the developed microsyringe pump.

References

1. Awtar, S., Parmar, G.: De sign of a large range xy nanopositioning system. J. Mech. Robot. **5**, 021008 (2013)
2. Cao, L., Mantell, S., Polla, D.: Design and simulation of an implantable medical drug delivery system using microelectromechanical systems technology. Sens. Actuators, A **94**, 117–125 (2001)
3. Chen, G., Han, X., Liu, G., Jiang, C., Zhao, Z.: An efficient multi-objective optimization method for black-box functions using sequential approximate technique. Appl. Soft Comput. **12**, 14–27 (2012)
4. He, Y., Lee, H.: Liquid-phase microextraction in a single drop of organic solvent by using a conventional microsyringe. Anal. Chem. **69**, 4634–4640 (1997)
5. Hodgkin, A., Keynes, R.: Experiments on the injection of substances into squid giant axons by means of a microsyringe. J. Physiol. **131**, 592 (1956)
6. Jaffe, B.: Piezoelectric Ceramics. Academic Press, London (1971)
7. Kan, J., Yang, Z., Peng, T., Cheng, G., Wu, B.: Design and test of a high-performance piezoelectric micropump for drug delivery. Sens. Actuators, A **121**(1), 156–161 (2005)
8. Liu, G., Yang, Z., Liu, J., Li, X., Wang, H., Zhao, T., Yang, X.: A low cost, high performance insulin delivery system based on pzt actuation. Microsyst. Technol. **20**(12), 2287–2294 (2014)
9. Liu, Y., Wu, K., Xu, D., Xu, Q.: Compliant constant-force micro-mechanism for enabling dual-stage motion. In: Proceedings of 2014 IEEE International Conference on Information and Automation (ICIA), pp. 1–6 (2014)
10. Nisar, A., Afzulpurkar, N., Mahaisavariya, B., Tuantranont, A.: MEMS-based micropumps in drug delivery and biomedical applications. Sens. Actuators, A **130**, 917–942 (2008)
11. Rondelet, J.C., Dupouy, J.M.: Syringe pump having continuous pressure monitoring and display (1994). US Patent 5295967 A
12. Sanderson, G.G., Massaglia, M., Palmer, M.J.: Syringe pump and the like for delivering medication (1990). US Patent 4976696 A
13. Wang, B., Chu, X., Li, E., Li, L.: Simulations and analysis of a piezoelectric micropump. Ultrasonics **44**, e643–e646 (2006)
14. Xu, Q.: New flexure parallel-kinematic micropositioning system with large workspace. IEEE Trans. Rob. **28**, 478–491 (2012)
15. Xu, Q., Li, Y.: Analytical modeling, optimization and testing of a compound bridge-type compliant displacement amplifier. Mech. Mach. Theory **46**, 183–200 (2011)
16. Xu, W., King, T.: Flexure hinges for piezoactuator displacement amplifiers: flexibility, accuracy, and stress considerations. Precis. Eng. **19**, 4–10 (1996)

Chapter 10
Visual Servo Control with Force Regulation for Microinjection

Abstract In this chapter, a microinjection system has been developed by integrating dual cameras and microforce sensor. Experimental study on microinjection of zebrafish embryos with visual servo control and force control has been conducted to improve the success rate of injection and survival rate of the injected cells. Experimental results demonstrate the effectiveness of the custom-built microinjection system.

10.1 Introduction

Genetic engineering is one of the hot topics in current scientific research. Many different kinds of materials, e.g., exogenous proteins, cDNA constructs, peptides, drugs, and particles, have been successfully delivered into cellulars via microinjection systems [21]. These systems can increase the efficiency of genetic engineering work and reduce the workload of researchers. In the literature, lots of new mechanisms [19], control algorithms [17], and sensors [10] have been proposed for microinjection systems.

Despite the various research results, there are only a few feedback-sensing techniques (including visual and force approaches) that can work at the micro-/nanoscale [12]. All of these techniques have been studied extensively in recent years [4, 9, 18]. As mentioned in [12], there are many advantages for visual feedback control in micro-/nanoscale operation. In particular, vision can provide the parameters of multi-objects in the same coordinate. Such approach can essentially reduce the error caused by coordinate transformation. Meanwhile, in the micro-/nanoscale, the degree of nonlinearity of the measurement results is very low, and it can still give accurate results if treated as a linear system. In the literature, Sun et al. achieved the positioning precision of ±16 nm for a nano-manipulator tip with template matching algorithm [13]. They controlled the multi-axis microactuator with the help of standard microscope CCD camera, and a control rate of 30 Hz is realized. Karimirad et al. presented a

Q. Xu, *Micromachines for Biological Micromanipulation*,
https://doi.org/10.1007/978-3-319-74621-0_10

vision-based force measurement method using an artificial neural network model [6]. They used discrete cosine transform to describe the deformation of the cell and predicted the injection force with this deformation. Their model gives a good prediction result. Huang et al. used chord midpoint Hough transform to detect the cell in an image [5]. The cells in the experiment are approximately circular, and their diameters can be obtained.

Even though the advantage of microvision is obvious, it is costly to obtain the depth information in microvision, which is its main disadvantage. So, most researches have applied the micro-visual servoing system under two-dimensional conditions without depth detection. Another disadvantage of visual servoing system is that the computation load is quite large. Specifically, the computational complexity in traditional template matching algorithm is $O(n^4)$, which cannot give a real-time solution. Thus, some improved template matching algorithms are presented [2, 20]. Thakar et al. reduced the computational delay by using the resized images [15]. However, their result also exhibits error which is produced by the resizing operation. Another research is chaotic imperialist competitive image matching approach [3], which is presented by Duan et al. Their experimental results confirm an obvious improvement. Moreover, the block sum pyramid algorithm as proposed in [7] uses the coarse to fine operation in an integrated image, which gives precise result and the computation speed is enhanced. Hence, the algorithm is employed in this chapter to achieve a better performance.

In this chapter, a new cell microinjection system with dual cameras is proposed to guide the cell injection operation. Unlike the existing recent researches about 3D visual servo system in micro-/nanoscale [11], the proposed system can offer the depth information more directly. Experimental result confirms the feasibility of this feedback method. In order to detect the injector and cells, template matching algorithm has been implemented. The effectiveness of the developed microinjection system has been confirmed by carrying out experimental study of zebrafish embryos microinjection.

10.2 Experimental Setup

A photograph of the experimental setup is illustrated in Fig. 10.1. The system is composed of three parts, i.e., vision unit, manipulation unit, and host computer. The vision unit consists of two microscopes with cameras (Cam A and Cam B). The camera parameters are shown in Table 10.1.

The manipulation unit contains an XYZ micromanipulator (model: MPC-385, from Sutter Instrument Inc.), a motorized high-speed XY stage (model: MLS203-2P, from Thorlabs Inc.), a picoliter microinjector (model: PLI-100A, from Warner Instruments Inc.), and a piezo-driven linear positioner. The high-precision XYZ micromanipulators are employed to control the position of the injector. The XYZ micromanipulator can travel in the maximum range of 25 mm with a speed of 5 mm/s and step resolution of 0.25 μm. The XY scanning stage holds a petri dish with V-shaped

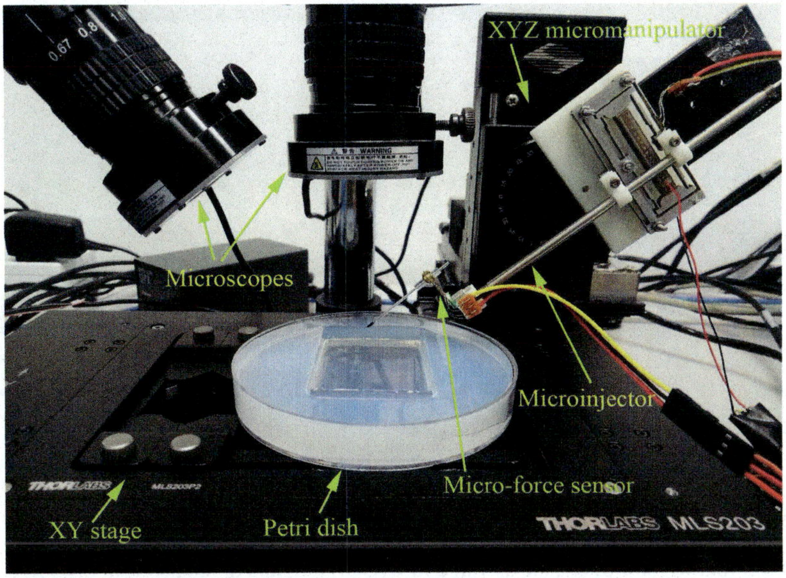

Fig. 10.1 Experimental setup for a custom-built cell microinjection system with dual cameras

Table 10.1 Main parameters of the microscopes with cameras

	Cam A	Cam B
Model	SK2700VS-B2	SK2700VS-B1
Lens magnification range	32× to 317×	38× to 270×
Installation angle relative to vertical direction	0°	47.1°
Resolution (pixel)	1600 × 1200	1600 × 1200
Frames per second	15	15

grooves fabricated of agarose gel, where the cells are placed. The maximum travel range of the XY scanning stage is 110 mm × 75 mm with the maximum velocity of 250 mm/s and absolute on-axis accuracy better than 3 μm. The XYZ micromanipulator and XY stage are connected to a host computer by RS-232 serial communication ports with 128000 baud rate. The picoliter microinjector can produce the injection pressure of 1.37–413 kPa and holding vacuum ranging from 0 to −0.75 kPa. The piezo-driven linear positioner is actuated by a piezoelectric actuator (model: EPA-104-230, from Piezo Systems, Inc.) with the maximum peak voltage of ±200 V, maximum peak current of ±200 mA, maximum peak power of 40 V at frequencies no more than 250 kHz. The piezoresistive microforce sensor is glued on a PVDF film, and a micropipette injector is mounted on the microforce sensor for cell injection purpose [16]. The system is controlled in real time by the host computer running LabVIEW software (from National Instruments Corp.).

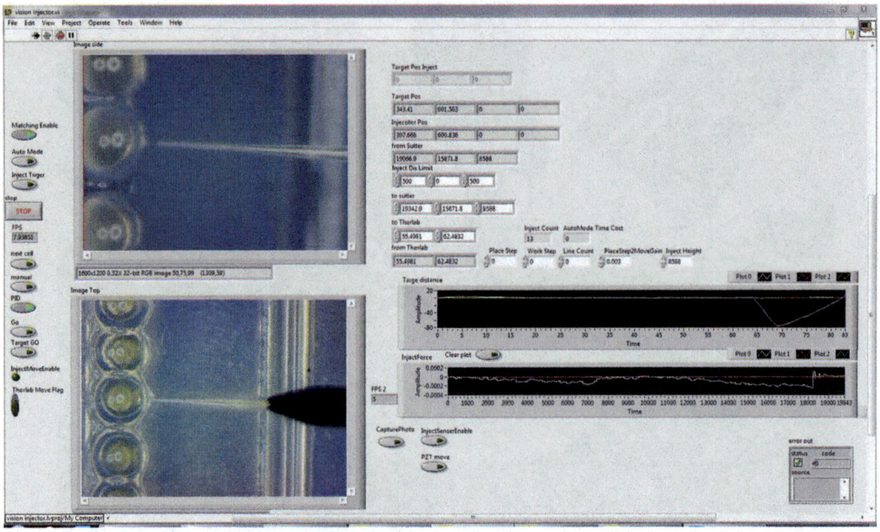

Fig. 10.2 Front window of graphic user interface (GUI) program developed with LabVIEW

The graphical user interface (GUI) of the program is shown in Fig. 10.2. The two image display windows show the captured images from Cam A and Cam B in real time. The “Target Distance” wave chart gives the PID control error at every process. The “Inject Force” wave chart shows the piercing force. The control parameters are shown at the top-right side of the GUI window, while control buttons are placed at the left and bottom-right sides for the convenience of operation.

By aligning the cell arrays on the V-shaped agarose gel grooves in the petri dish manually, the objective of the concerned microinjection is to pierce the cells and inject drug into the cells automatically by the microinjector with motion and force control. The development process of the cell microinjection system is presented in the following.

10.3 Image Processing Procedure

Once the images are acquired, image processing is conducted to identify the injector and cells. In this work, the template matching method is used to detect the injector and cells, respectively. Two representative images (for the injector and cell, respectively) are saved as the templates. At each available position, the acquired source images are compared with the two template images. With the comparison result, one or more positions where they are quite similar can be obtained.

10.3.1 Detection of the Injector

Let $I(x, y)$ represent the pixel value of the source image at location (x, y). $T(x, y)$ expresses the pixel value of the template image, and $R(x, y)$ indicates the comparison result. The simplest comparison equation can be given as follows [8].

$$R(x, y) = \sum_{i,j} [I(x + i, y + j) - T(i, j)]^2 \tag{10.1}$$

where x and y represent the relative positions of the template image in the source image. With the help of this equation, the computation solution can be obtained.

If the size of the template image is $m_t \times n_t$ and the size of the target image is $m_i \times n_i$, the computation solution will be a $(m_i - m_t)$ by $(n_i - n_t)$ matrix (mostly, it is expressed as a gray solution image). In this solution, the lower the pixel value is, the higher the template image matches. So, if the threshold value is set, several locations can be derived, which represent the matching solutions.

Another problem in the template matching algorithm is the energy. That is, if the energy meets $\sum I(x, y) \neq \sum T(x, y)$ in any of the matching region, the matching algorithm (10.1) will give an incorrect solution. Normalized cross-correlation can overcome this problem by normalizing both of the template image and the matching region of the source image into the same interval, so that their energies are equal [1]. One of the widely used normalizing equation is given as:

$$R'(x, y) = \frac{\sum_{i,j} [I(x + i, y + j) - \overline{I}][T(i, j) - \overline{T}]}{\sqrt{\sum_{i,j} [I(x + i, y + j) - \overline{I}]^2 \sum_{i,j} [T(i, j) - \overline{T}]^2}} \tag{10.2}$$

where $\overline{I}$ and $\overline{T}$ are the average values in the matching region of the source image and template image, respectively. This equation will give a result of 1 if the template equals the source image completely. The more different they are, the lower value this equation obtains.

Traditional template matching algorithm (say, cross-correlation) costs too much computation load. Hence, an improved algorithm is necessary. In this chapter, the block sum pyramid algorithm is employed. The principle of this algorithm is to reduce the size of the template image and target image with proportions, and then repeat this operation several times. The smallest image group is matched first, because of the low size of them. This matching result costs a shorter time and produces a relatively low-precision solution. One group with higher size will be matched in the region as given in the previous matching result. These operations are repeated until the group with the highest size is matched, and the most precision solution can be obtained [7]. This is the so-called coarse-to-fine process. The low-size image gives the approximate position and reduces the matching region of the high-resolution images. If each smaller image is placed over a bigger one, it looks like a pyramid, which is the origin of its name.

In majority of current researches, the template matching algorithm is employed to match the location of the target. For example, the work [13] matched the top section of the microactuator, and Nelson et al. guided the locating of the injector with its peak point [14]. Whereas in the microinjection system, the peak of the inject pipette will pierce into the cell. It is hard to position these objects, even though the injected cell is transparent. Therefore, in this work, the template image is not set as the image of the peak of the injecting pipette, but its middle section. Meanwhile, as shown in Fig. 10.3, the midsection of the injector is marked as black to improve the matching reliability. The location of the pipette peak can be computed by the given location of its midsection, which is constant during the microinjection procedure.

If the detected position is (x_m, y_m, z_m), and detected angles of its midsection are θ_1 in Cam A and θ_2 in Cam B, the pipette peak position (x_p, y_p, z_p) can be given as follows.

$$\begin{cases} x_p = x_m - d\cos(\theta_1) \\ y_p = y_m - d\sin(\theta_1) \\ z_p = z_m - d\sin(\theta_2) \end{cases} \tag{10.3}$$

where d represents the distance between the peak and the midsection of the pipette as shown in Fig. 10.4a.

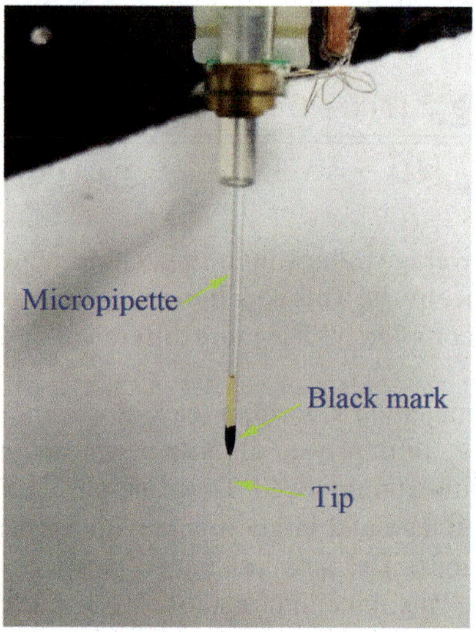

Fig. 10.3 Photograph of the micropipette injector with black mark

10.3.2 Detection of the Cells

In order to locate the target cells, the template matching algorithm is also employed to the cell detection. Figure 10.4b shows the chosen cell template which is used in the cell microinjection experiment.

Using the selected two templates for the injector and cell, the template matching results are illustrated in Fig. 10.5. It is observed that the algorithm can detect both features successfully. Specifically, the green rectangle A is the region of interest (ROI) for the cell matching algorithm, which only covers a part of the whole image to ensure that no more than one cell will appear in the matching area. The green rectangle B represents the ROI for the injector matching process. It covers the whole image to reduce the possibility of injector loss in the field of view, because the injector moves in a large range during the injection procedure.

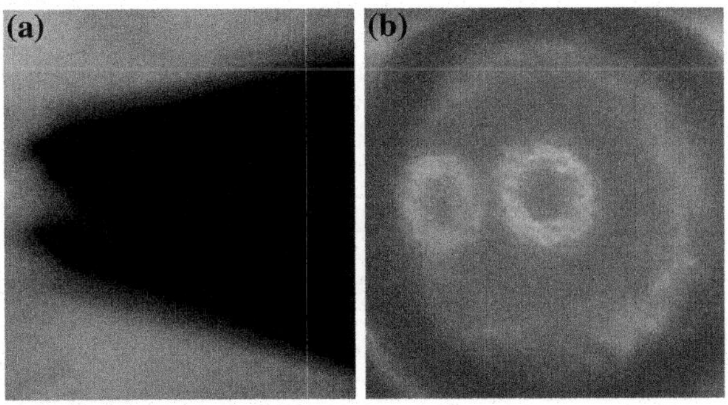

Fig. 10.4 Selected image templates for **a** injector and **b** cell

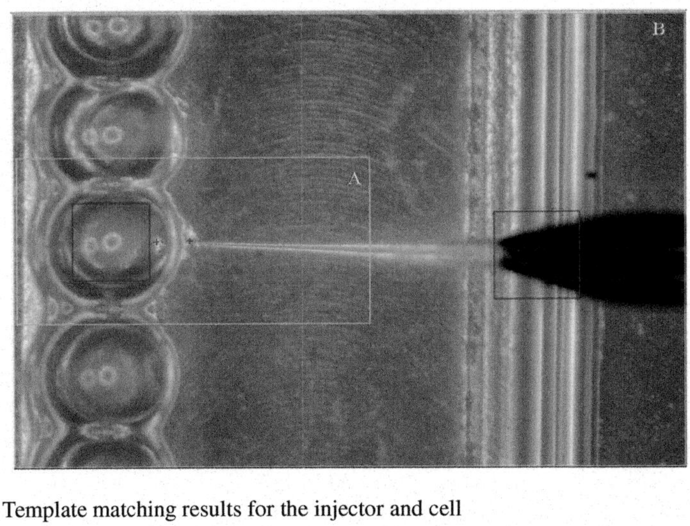

Fig. 10.5 Template matching results for the injector and cell

10.4 Control Scheme Design

The overall control scheme is illustrated in the block diagram as shown in Fig. 10.6. The control sequences are described as follows.

1. Cell searching: Move the uninjected cell to the center of screen for injecting (see more details later).
2. Injector alignment: Guide the injector to touch the cell with computer vision and PID control algorithm for the XY stage.

Fig. 10.6 Block diagram of overall control scheme

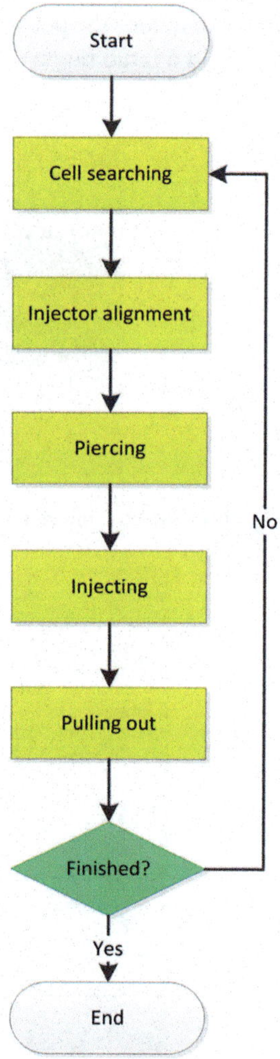

3. Piercing: Enable the operation of piezo-driven linear positioner. The force sensor provides the piercing force feedback, so that the piercing force is controllable (see more details later).
4. Injecting: Enable the injector signal. The drug will be injected into the pierced cell.
5. Pulling out: After the drug is injected, the injector will be pulled out of the cell to execute the subsequent injection process.

10.4.1 Cell Searching Process

In the cell searching process, the petri dish is translated along a planned path as shown in Fig. 10.7. Once a cell is injected, the petri dish will continue its movement to search for the next cell. The detailed searching process is shown in Fig. 10.8, which is outlined below.

1. Move injector back 1000 μm: The injector is moved backward with a specified distance to protect the injector itself and the cell during the subsequent petri dish movement.
2. Move the precious cell out of view: With this step, the template matching algorithm will not wrongly take the previous cell as new ones.
3. Move forward or backward: Move the petri dish forward (in odd columns) or backward (in even columns) with a given shift distance.
4. Move cell in view: Matching the ROI of the input image, continue moving the petri dish if no cell is found. Else, go to the next step if a cell is found.

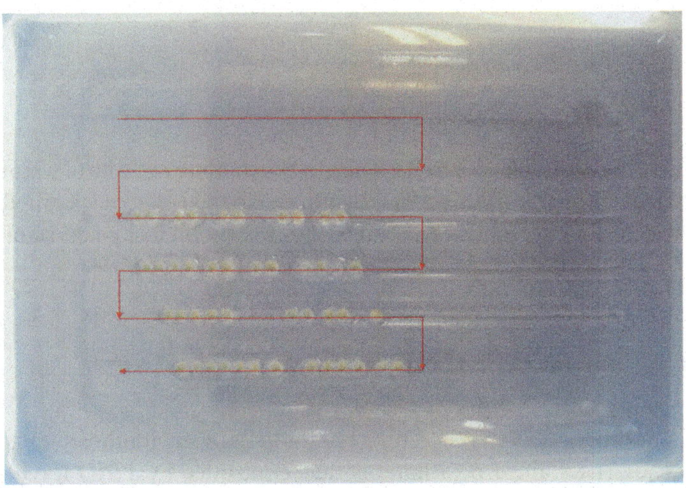

Fig. 10.7 Planned cell searching path on a petri dish

Fig. 10.8 Control block diagram of cell searching process

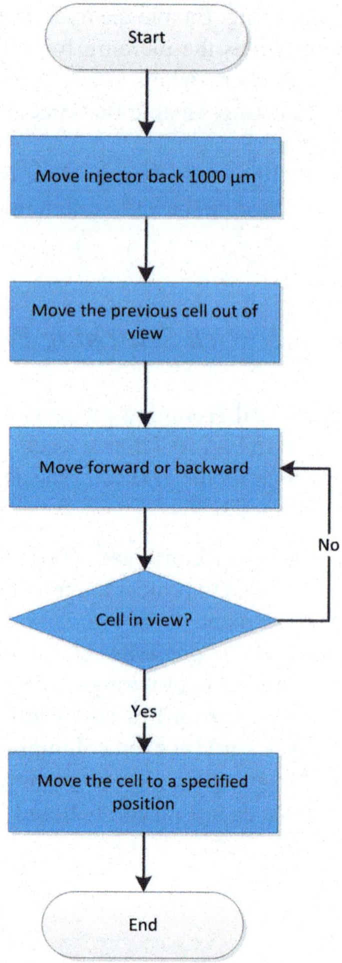

5. Move cell to a specified position: The PID visual servo control algorithm and template matching algorithm are applied to move the cell to a specified position. Then, the injector can contact the cell with highest efficiency and stability.

10.4.2 Cell Piercing Process

In the piercing process, a reference force signal (i.e., a sawtooth wave) is assigned first. The piezo-driven linear positioner is then controlled by a PID force control to ensure that the piercing force follows the preset force trajectory precisely. Meanwhile, if the force derivative gives a signal which is larger than a threshold, it indicates that the injector has pierced into cell membrane. Then, the force control algorithm will stop and the drug is injected into the cell. The cell piercing process is depicted in Fig. 10.9.

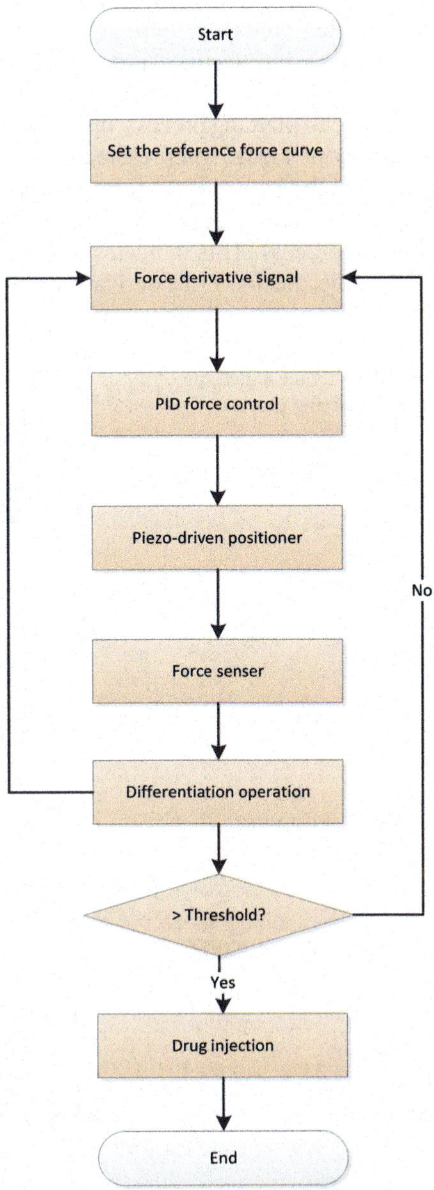

Fig. 10.9 Control block diagram of cell piercing process

10.5 Experimental Results

In the experiment, multiple cells in different arrays are injected one by one. With the developed LabVIEW program, the cells can be injected with two control modes, i.e., automatic mode and semi-automatic mode. In the automatic mode, the system will search and inject cells without any human intervention. This mode will finish all of the injection processes with an average time of 6 s per cell. In the semi-automatic mode, every intermediate process of system can be manually controlled and the sub-steps can be automatically completed.

In the cell piercing process, the piercing force signals are acquired with a sampling frequency of 1 kHz. A representative of the force signal is shown in Fig. 10.10. The first-order derivative of the force signal is given in Fig. 10.11.

It is observed that there is a dramatic drop in the force signal at the end of the cell piercing process. This indicates that the cell membrane is pierced by the injector. Meanwhile, the first-order derivative of the force signal gives a high peak value, which

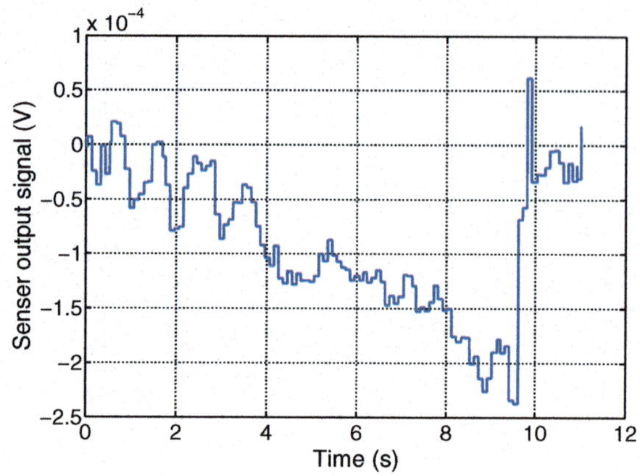

Fig. 10.10 Force signal in the cell piercing process

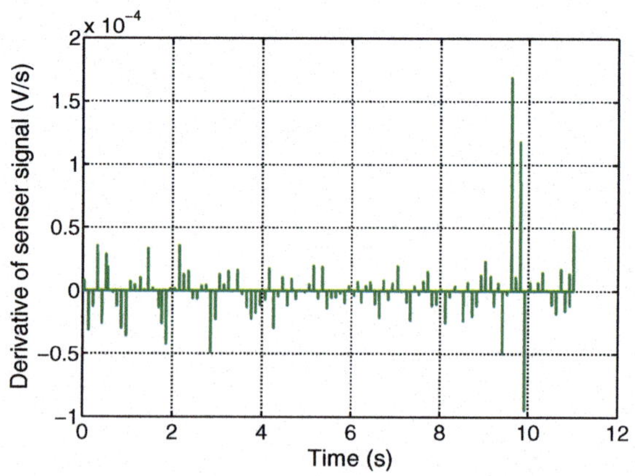

Fig. 10.11 First-order derivative of the force signal in the cell piercing process

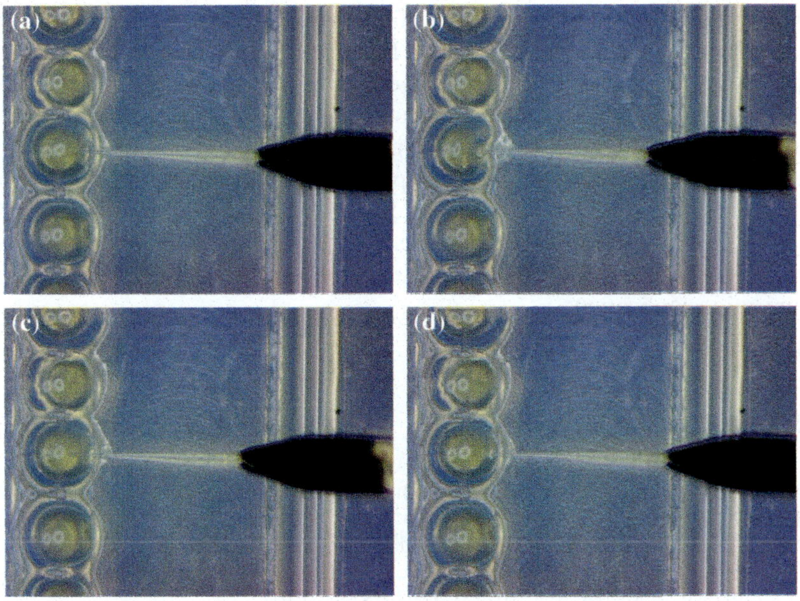

Fig. 10.12 Photograph sequences of the cell piercing process for zebrafish embryos

serves as an instruction to inject the drug into the cell. Moreover, the sequences of this piercing process are illustrated by the captured photographs as shown in Fig. 10.12.

Recently, the developed automated robotic cell microinjection system has been applied to gene knockout of zebrafish embryos. The experimental results of the automated microinjection have been compared with those of manual microinjection. Results indicate clearly that the developed microinjection system enables a higher survival rate and better consistency for the injected cells.

In the future work, the performance of the developed devices will be further improved. More promising applications will be carried out by the automated micro-robotic system with various advanced micromachines to facilitate the multidisciplinary researches for clinical and industrial applications in biomedical and health sciences.

10.6 Conclusion

The experimental results show that the developed microinjection system can achieve an automated batch cell injection with a speed of 6 s per cell. This speed is close to that of a skilled operator, but the system saves more manpower and produces high success rate and high consistency in injection. When a large number of cells need to be injected, the developed system can continue working, while the human operators need rest during the manipulation. With a high prediction accuracy for the position of injector tip, the system can precisely recognize the injector tip's location even though it is immersed into the cell and cannot be observed directly. Therefore, the microinjection system provides a reliable injection of drug into a specified position of the cells. Moreover, the force control approach ensures that the cell suffers from the lowest injury during the piercing process, and any errors in this process can be easily identified in advance in the microinjection tasks. In the future work, high-speed cameras will be employed to improve the injection speed for clinical and industrial microinjection applications.

References

1. Briechle, K., Hanebeck, U.D.: Template matching using fast normalized cross correlation. Proc. SPIE **4387**, 95–102 (2001)
2. Cuevas, E., Osuna, V., Oliva, D.: Template Matching, pp. 65–93. Springer, Cham (2017)
3. Duan, H., Xu, C., Liu, S., Shao, S.: Template matching using chaotic imperialist competitive algorithm. Pattern Recogn. Lett. **31**(13), 1868–1875 (2010)
4. Huang, H., Sun, D., Mills, J.K., Li, W.J., Cheng, S.H.: Visual-based impedance control of out-of-plane cell injection systems. IEEE Trans. Autom. Sci. Eng. **6**(3), 565–571 (2009)
5. Huang, H.B., Sun, D., Mills, J.K., Cheng, S.H.: Robotic cell injection system with position and force control: toward automatic batch biomanipulation. IEEE Trans. Robot. **25**(3), 727–737 (2009)
6. Karimirad, F., Chauhan, S., Shirinzadeh, B.: Vision-based force measurement using neural networks for biological cell microinjection. J. Biomech. **47**(5), 1157–1163 (2014)
7. Lee, C.H., Chen, L.H.: A fast motion estimation algorithm based on the block sum pyramid. IEEE Trans. Image Process. **6**(11), 1587–1591 (1997)
8. Lewis, J.P.: Fast template matching. Vision Interface **95**, 15–19 (1995)
9. Liu, J., Zhang, Z., Wang, X., Liu, H., Zhao, Q., Zhou, C., Tan, M., Pu, H., Xie, S., Sun, Y.: Automated robotic measurement of 3-D cell morphologies. IEEE Robot. Autom. Let. **2**(2), 499–505 (2017)
10. Pillarisetti, A., Anjum, W., Desai, J.P., Friedman, G., Brooks, A.D.: Force feedback interface for cell injection. In: Proceedings of the First Joint Eurohaptics Conference and Symposium on Haptic Interfaces for Virtual Environment and Teleoperator Systems, pp. 391–400 (2005)
11. Sano, T., Yamamoto, H.: A visual servo system for microinjection under stereoscopic microscope. In: Proceedings of 9th IFAC Symposium on Artificial Intelligence in Real Time Control (AIRTC 2000), vol. 33, No. 28, pp. 101–106 (2000)
12. Shen, Y., Fukuda, T.: State of the art: micro-nanorobotic manipulation in single cell analysis. Robot. Biomim. **1**(1), 21 (2014)
13. Sun, Y., Greminger, M.A., Nelson, B.J.: Nanopositioning of a multi-axis microactuator using visual servoing. J. Micromechtron. **2**(2), 141–155 (2002)
14. Sun, Y., Nelson, B.J.: Biological cell injection using an autonomous microrobotic system. Int. J. Robot. Res. **21**(10–11), 861–868 (2002)
15. Thakar, K., Kapadia, D., Natali, F., Sarvaiya, J.: Implementation and analysis of template matching for image registration on devkit-8500d. Optik—Int. J. Light Electron Opt. **130**, 935–944 (2017)
16. Wang, G., Xu, Q.: Design and development of a piezo-driven microinjection system with force feedback. Adv. Robot. **31**(23–24), 1349–1359 (2017). https://doi.org/10.1080/01691864.2017.1362996

17. Wang, G., Xu, Q.: Design and precision position/force control of a piezo-driven microinjection system. IEEE/ASME Trans. Mechatron. **22**(4), 1744–1754 (2017)
18. Wang, W., Liu, X., Gelinas, D., Ciruna, B., Sun, Y.: A fully automated robotic system for microinjection of zebrafish embryos. PLoS One **2**(9), e862 (2007)
19. Xu, Q.: Design and development of a compact flexure-based xy precision positioning system with centimeter range. IEEE Trans. Ind. Electron. **61**(2), 893–903 (2014)
20. Zhang, S., Zhou, Y.: Template matching using grey wolf optimizer with lateral inhibition. Optik—Int. J. Light Electron Opt. **130**, 1229–1243 (2017)
21. Zhang, Y., Yu, L.C.: Single-cell microinjection technology in cell biology. Bioessays **30**(6), 606–610 (2008)

Index

Q. Xu, *Micromachines for Biological Micromanipulation*,
https://doi.org/10.1007/978-3-319-74621-0

Printed by Printforce, the Netherlands